T0338457

CAMBRIDGE TRACTS IN MATHEMATICS

General Editors

203 A Primer on the Dirichlet Space

CAMBRIDGE TRACTS IN MATHEMATICS

A complete list of books in the series can be found at www.cambridge.org/mathematics.
Recent titles include the following:

169. Quantum Stochastic Processes and Noncommutative Geometry. By K. B. SINHA and D. GOSWAMI
170. Polynomials and Vanishing Cycles. By M. TIBĂR
171. Orbifolds and Stringy Topology. By A. ADEM, J. LEIDA, and Y. RUAN
172. Rigid Cohomology. By B. LE STUM
173. Enumeration of Finite Groups. By S. R. BLACKBURN, P. M. NEUMANN, and G. VENKATARAMAN
174. Forcing Idealized. By J. ZAPLETAL
175. The Large Sieve and its Applications. By E. KOWALSKI
176. The Monster Group and Majorana Involutions. By A. A. IVANOV
177. A Higher-Dimensional Sieve Method. By H. G. DIAMOND, H. HALBERSTAM, and W. F. GALWAY
178. Analysis in Positive Characteristic. By A. N. KOCHUBEI
179. Dynamics of Linear Operators. By F. BAYART and É. MATHERON
180. Synthetic Geometry of Manifolds. By A. KOCK
181. Totally Positive Matrices. By A. PINKUS
182. Nonlinear Markov Processes and Kinetic Equations. By V. N. KOLOKOLTSOV
183. Period Domains over Finite and p-adic Fields. By J.-F. DAT, S. ORLIK, and M. RAPOPORT
184. Algebraic Theories. By J. ADÁMEK, J. ROSICKÝ, and E. M. VITALE
185. Rigidity in Higher Rank Abelian Group Actions I: Introduction and Cocycle Problem. By A. KATOK and V. NIŢICĂ
186. Dimensions, Embeddings, and Attractors. By J. C. ROBINSON
187. Convexity: An Analytic Viewpoint. By B. SIMON
188. Modern Approaches to the Invariant Subspace Problem. By I. CHALENDAR and J. R. PARTINGTON
189. Nonlinear Perron–Frobenius Theory. By B. LEMMENS and R. NUSSBAUM
190. Jordan Structures in Geometry and Analysis. By C.-H. CHU
191. Malliavin Calculus for Lévy Processes and Infinite-Dimensional Brownian Motion. By H. OSSWALD
192. Normal Approximations with Malliavin Calculus. By I. NOURDIN and G. PECCATI
193. Distribution Modulo One and Diophantine Approximation. By Y. BUGEAUD
194. Mathematics of Two-Dimensional Turbulence. By S. KUKSIN and A. SHIRIKYAN
195. A Universal Construction for Groups Acting Freely on Real Trees. By I. CHISWELL and T. MÜLLER
196. The Theory of Hardy's Z-Function. By A. IVIĆ
197. Induced Representations of Locally Compact Groups. By E. KANIUTH and K. F. TAYLOR
198. Topics in Critical Point Theory. By K. PERERA and M. SCHECHTER
199. Combinatorics of Minuscule Representations. By R. M. GREEN
200. Singularities of the Minimal Model Program. By J. KOLLÁR
201. Coherence in Three-Dimensional Category Theory. By N. GURSKI
202. Canonical Ramsey Theory on Polish Spaces. By V. KANOVEI, M. SABOK, and J. ZAPLETAL
203. A Primer on the Dirichlet Space. By O. EL-FALLAH, K. KELLAY, J. MASHREGHI, and T. RANSFORD

A Primer on the Dirichlet Space

OMAR EL-FALLAH
Université Mohammed V-Agdal, Rabat, Morocco

KARIM KELLAY
Université Bordeaux 1, Bordeaux, France

JAVAD MASHREGHI
Université Laval, Québec, Canada

THOMAS RANSFORD
Université Laval, Québec, Canada

CAMBRIDGE
UNIVERSITY PRESS

University Printing House, Cambridge CB2 8BS, United Kingdom

Cambridge University Press is part of the University of Cambridge.

It furthers the University's mission by disseminating knowledge in the pursuit of education, learning and research at the highest international levels of excellence.

www.cambridge.org
Information on this title: www.cambridge.org/9781107047525

First published 2014

A catalogue record for this publication is available from the British Library

ISBN 978-1-107-04752-5 Hardback

To:
Zaïnab and Hiba,
Anna and Maël,
Dorsa, Parisa and Golsa,
Julian and Étienne

Contents

Preface

The three classical Hilbert spaces of holomorphic functions in the unit disk are the Hardy, Bergman and Dirichlet spaces. There are several excellent texts covering the Hardy space and the Bergman space. However, to the best of our knowledge, up to now there has been no book devoted to the Dirichlet space. When we began our respective researches into the Dirichlet space, we found ourselves handicapped by the fact that the necessary background information was scattered around the literature, sometimes contained in articles that were difficult to follow. For this reason we began to think about writing an introduction that would be suitable for researchers and graduate students seeking a solid background in the subject. The more we learned about this topic, the more we became convinced that it contains many beautiful ideas that deserve a systematic exposition.

The name Dirichlet space derives from its definition in terms of the so-called Dirichlet integral, arising in Dirichlet's method for solving Laplace's equation (sometimes called the Dirichlet principle). As far as we can determine, the first appearance of the Dirichlet space under that name dates back to two articles of Beurling and Deny in 1958 and 1959, but in fact the notion existed and had been studied at least since Beurling's thesis, which was published in 1933 and written even a little earlier. In the years that followed, Beurling and Carleson laid the foundations of the theory and, after their pioneering work, many other distinguished mathematicians made important contributions.

Why study the Dirichlet space? Here are a few reasons.

(1) The Hardy space corresponds to ℓ^2, the Hilbert space of square-summable sequences. One of the main advantages of thinking of it as a function space is that the shift operator on ℓ^2 becomes simply multiplication by z. If one is interested in weighted shifts on ℓ^2, which are very important in operator theory, then one should consider multiplication by z on a weighted function

space. The two most basic non-constant weights lead one immediately to the Dirichlet and Bergman spaces.

(2) The Dirichlet integral of a holomorphic function f has a very natural geometric interpretation. It is exactly the area of the image of f, counted according to multiplicity. Seen this way, it is obviously invariant under precomposition with every Möbius automorphism of the unit disk. It is a remarkable fact that this Möbius-invariance property characterizes the Dirichlet space among all Hilbert function spaces on the disk.

(3) The Dirichlet space is closely related to logarithmic potential theory. In particular, the notions of energy and logarithmic capacity play a prominent role in the theory. This reflects Beurling's vision of the subject, and yields interesting interactions with physics.

(4) The Dirichlet integral is the motivating example of the abstract notion of a Dirichlet form, first introduced by Beurling and Deny in the articles mentioned above. Dirichlet forms have become a fundamental tool in probability and semigroup theory (though this is not an aspect that will be developed in this book).

(5) From many points of view, the Dirichlet space is a borderline case. For example, it is very nearly an algebra, but not quite. This borderline nature makes it an interesting and challenging example of a function space. Many important questions remain unsolved, and the Dirichlet space is still an active area of research.

What is in the book? To get a quick idea, imagine being presented with a function space on the unit disk. Several standard questions naturally arise. For example:

- What can be said about the boundary behavior of functions in the space?
- Are there simple characterizations of zero sets and uniqueness sets?
- What can we say about interpolation?
- Is the space an algebra? If not, then what are the multipliers?
- How rich is the operator theory on this function space? For example, can we classify the shift-invariant subspaces? Which functions are cyclic?

In the case of the Hardy space, the answers to all these questions are well known and important. By contrast, in the Dirichlet space, some of the questions have been only partially answered, and even where the complete answers are known, they are more subtle. This is the subject of this book.

Perhaps it is also worth mentioning what is not in the book. As it is meant to be a primer, we do not pretend to give an exhaustive treatment of the subject, and certain topics, such as interpolating sequences and the corona problem, have been omitted completely (with much regret). We have decided to restrict ourselves to the classical Dirichlet space, treating other variants such as weighted Dirichlet spaces when they contribute directly to understanding the classical case.

The prerequisites are a knowledge of standard complex analysis, measure theory and functional analysis. Also, we have taken for granted a certain familiarity with Hardy spaces, the necessary background being summarized briefly in an appendix. We do however develop the notion of logarithmic capacity *ab initio*, since it turns up throughout the book.

There are exercises at the end of most of the sections, ranging from routine calculations to barely disguised theorems. We have tried our best to attribute results correctly, in notes at the end of each chapter. However, history is sometimes complicated, and we apologize if we have fallen short of our aim.

In the course of writing the book, we have benefitted from discussions with many mathematicians. In particular, we thank Alexandru Aleman, Nicola Arcozzi, Sasha Borichev, Håkan Hedenmalm, Stefan Richter, Bill Ross, Kristian Seip and Andrew Wynn. Also we thank Jérémie Rostand for his help with the illustrations. We are grateful to Roger Astley and his colleagues at Cambridge University Press for their advice and encouragement. Part of this book was written at the CIRM (Luminy), and we express our gratitude to the CIRM for its hospitality. We gratefully acknowledge the financial support of the following granting bodies: CNRST and the Hassan II Academy of Science and Technology (OE), PICS-CNRS (KK), NSERC (JM and TR) and the Canada research chairs program (TR). Of course, we owe a huge debt of gratitude to our spouses, Salma, Nathalie, Shahzad and Line, for supporting us and putting up with us while the book was being written. Last, but not least, we thank our children for constantly reminding us that there are things even more important than mathematics. We dedicate the book to them.

1

Basic notions

In this introductory chapter, we introduce the Dirichlet space. We develop its elementary properties, at the same time pointing the way to the deeper results to be treated in the subsequent chapters.

1.1 The Dirichlet space

Let us begin with the fundamental definition. In what follows, $\mathbb{D}$ denotes the open unit disk, $\mathrm{Hol}(\mathbb{D})$ is the set of all functions holomorphic in $\mathbb{D}$, and dA denotes the area Lebesgue measure on the complex plane $\mathbb{C}$.

Definition 1.1.1 Given $f \in \mathrm{Hol}(\mathbb{D})$, the *Dirichlet integral* of f is defined by

$$\mathcal{D}(f) := \frac{1}{\pi} \int_{\mathbb{D}} |f'(z)|^2 \, dA(z).$$

The *Dirichlet space* $\mathcal{D}$ is the vector space of $f \in \mathrm{Hol}(\mathbb{D})$ such that $\mathcal{D}(f) < \infty$.

Clearly $\mathcal{D}$ contains all the polynomials and, more generally, all functions holomorphic on $\overline{\mathbb{D}}$ such that f' is bounded on $\mathbb{D}$. We shall see that it also contains many other interesting functions.

Our first result is a formula for $\mathcal{D}(f)$ in terms of the Taylor coefficients of f.

Theorem 1.1.2 *Let $f \in \mathrm{Hol}(\mathbb{D})$, say $f(z) = \sum_{k \geq 0} a_k z^k$. Then*

$$\mathcal{D}(f) = \sum_{k \geq 1} k |a_k|^2. \tag{1.1}$$

Proof Writing the area integral in polar coordinates, we have

$$\mathcal{D}(f) = \frac{1}{\pi} \int_{\mathbb{D}} \Big| \sum_{k \geq 1} k a_k z^{k-1} \Big|^2 dA(z) = \frac{1}{\pi} \int_0^1 \int_0^{2\pi} \Big| \sum_{k \geq 1} k a_k r^{k-1} e^{i(k-1)\theta} \Big|^2 d\theta \, r \, dr.$$

By Parseval's formula, for each $r \in (0,1)$,

$$\frac{1}{2\pi}\int_0^{2\pi}\Big|\sum_{k\geq 1} ka_k r^{k-1}e^{i(k-1)\theta}\Big|^2\,d\theta = \sum_{k\geq 1} k^2|a_k|^2 r^{2k-2}.$$

Hence

$$\mathcal{D}(f) = 2\int_0^1 \sum_{k\geq 1} k^2|a_k|^2 r^{2k-1}\,dr = \sum_{k\geq 1} k|a_k|^2. \qquad \square$$

From this, we derive the following simple but useful observation. Recall that the Hardy space H^2 consists of those holomorphic functions $f(z) = \sum_{k\geq 0} a_k z^k$ such that $\|f\|_{H^2}^2 := \sum_{k\geq 0}|a_k|^2 < \infty$.

Corollary 1.1.3 *The Dirichlet space $\mathcal{D}$ is contained in the Hardy space H^2.*

Proof This follows immediately from the theorem, together with the obvious fact that $\sum_{k\geq 0} k|a_k|^2 < \infty$ implies $\sum_{k\geq 0}|a_k|^2 < \infty$. $\square$

We shall make frequent use of this inclusion, exploiting the many known properties of the Hardy space H^2. A summary of the properties that we shall need can be found in Appendix A.

Notice that $\mathcal{D}$ is dense in H^2, because it contains the polynomials. As it is obviously a proper subspace of H^2, it is not closed in H^2, so it is not a Hilbert space with respect to $\|\cdot\|_{H^2}$. We now endow it with a norm, making it a Hilbert space in its own right.

For $f, g \in \mathcal{D}$, define

$$\mathcal{D}(f,g) := \frac{1}{\pi}\int_{\mathbb{D}} f'(z)\overline{g'(z)}\,dA(z).$$

This is a semi-inner product, and clearly $\mathcal{D}(f,f) = \mathcal{D}(f)$. In particular, $\mathcal{D}(f)^{1/2}$ is a semi-norm on $\mathcal{D}$. It is not quite a norm, since $\mathcal{D}(f) = 0$ whenever f is a constant. To get round this, we define

$$\langle f,g\rangle_{\mathcal{D}} := \langle f,g\rangle_{H^2} + \mathcal{D}(f,g) \qquad (f,g\in\mathcal{D}),$$

where $\langle\cdot,\cdot\rangle_{H^2}$ is the usual inner product on H^2. This gives a genuine inner product on $\mathcal{D}$, and the corresponding norm $\|\cdot\|_{\mathcal{D}}$ is given by

$$\|f\|_{\mathcal{D}}^2 = \|f\|_{H^2}^2 + \mathcal{D}(f) \qquad (f\in\mathcal{D}).$$

Theorem 1.1.4 *The Dirichlet space $\mathcal{D}$ is a Hilbert space with respect to the norm $\|\cdot\|_{\mathcal{D}}$.*

Proof Writing $f(z) = \sum_{k\geq 0} a_k z^k$, we have $\|f\|_{\mathcal{D}}^2 = \sum_{k\geq 0}(k+1)|a_k|^2$. Thus the map $f \mapsto ((k+1)^{1/2}a_k)_{k\geq 0}$ is an isometry of $\mathcal{D}$ onto ℓ^2, the space of square summable sequences. As ℓ^2 is a Hilbert space, so too is $\mathcal{D}$. $\square$

This is not the only way to make $\mathcal{D}$ a Hilbert space. Another common choice is to take $\|f\|^2 := |f(0)|^2 + \mathcal{D}(f)$. This also gives a Hilbert-space norm on $\mathcal{D}$, equivalent to $\|\cdot\|_{\mathcal{D}}$. However, unless stated otherwise, we shall always assume that $\mathcal{D}$ carries the norm $\|\cdot\|_{\mathcal{D}}$.

Exercises 1.1

1. Let $f \in \mathcal{D}$. Prove that $\|f\|_{\mathcal{D}}^2 \leq |f(0)|^2 + 2\mathcal{D}(f)$.
2. Let $f \in \mathcal{D}$, say $f(z) = \sum_{k\geq 0} a_k z^k$.

   (i) Let $s_n(z) := \sum_{k=0}^n a_k z^k$. Prove that $\|s_n\|_{\mathcal{D}} \leq \|f\|_{\mathcal{D}}$ for all $n \geq 0$, and that $\|s_n - f\|_{\mathcal{D}} \to 0$ as $n \to \infty$.

   (ii) Let $f_r(z) := f(rz)$. Prove that $\|f_r\|_{\mathcal{D}} \leq \|f\|_{\mathcal{D}}$ for all $0 < r < 1$, and that $\|f_r - f\|_{\mathcal{D}} \to 0$ as $r \to 1^-$.

3. Let $(f_n)_{n\geq 1} \in \mathcal{D}$. Show that, if $f_n \to f$ locally uniformly on $\mathbb{D}$, then we have $\mathcal{D}(f) \leq \liminf_{n\to\infty} \mathcal{D}(f_n)$. Deduce that, if $\sup_n \mathcal{D}(f_n) < \infty$, then $f \in \mathcal{D}$.
4. The *analytic Wiener algebra* is $W^+ := \{\sum_{k\geq 0} a_k z^k : \sum_{k\geq 0} |a_k| < \infty\}$. Give examples to show that $\mathcal{D} \not\subset W^+$ and that $W^+ \not\subset \mathcal{D}$.
5. The *disk algebra* $A(\mathbb{D})$ is the set of continuous functions $f : \overline{\mathbb{D}} \to \mathbb{C}$ that are holomorphic in $\mathbb{D}$. Give examples to show that $\mathcal{D} \not\subset A(\mathbb{D})$ and that $A(\mathbb{D}) \not\subset \mathcal{D}$.
6. (This exercise and the next make use of the Hardy spaces H^p for general p. For a summary of their properties, see Appendix A.) Let $f \in \operatorname{Hol}(\mathbb{D})$.

   (i) Show that, if $f' \in H^2$, then $f \in \mathcal{D}$.

   (ii) Show that, if $f' \in H^1$, then $f \in \mathcal{D}$. [This is somewhat harder. Use Hardy's inequality (Theorem A.1.9) to show that, if $f' \in H^1$, then the Taylor coefficients a_k of f satisfy $\sum_k |a_k| < \infty$ and $a_k = O(1/k)$.]

   (iii) Let $f(z) := \sum_{k\geq 1} z^k/k$. Show that $f' \in H^p$ for all $p < 1$ but $f \notin \mathcal{D}$.

7. We know that $\mathcal{D} \subset H^2$. Show that in fact $\mathcal{D} \subset \bigcap_{p<\infty} H^p$, as follows. Fix $f \in \mathcal{D}$ and $p \in (2, \infty)$.

   (i) Let $q \in (1, 2)$ be chosen so that $1/p + 1/q = 1$. Use Hölder's inequality to show that the Taylor coefficients a_k of f satisfy

$$\sum_k |a_k|^q \leq \Big(\sum_k \frac{1}{k^{q/(2-q)}}\Big)^{(2-q)/2} \Big(\sum_k k|a_k|^2\Big)^{q/2} < \infty.$$

   (ii) Use the Hausdorff–Young inequality to deduce that $f \in H^p$.

1.2 Reproducing kernels

We begin this section with a simple observation.

Theorem 1.2.1 *If $f \in \mathcal{D}$, then*

$$|f(z) - f(0)|^2 \leq \mathcal{D}(f) \log\Big(\frac{1}{1 - |z|^2}\Big) \qquad (z \in \mathbb{D}).$$

Proof Write $f(z) = \sum_{k \geq 0} a_k z^k$. For each $z \in \mathbb{D}$, the Cauchy–Schwarz inequality gives

$$|f(z) - f(0)|^2 = \Big|\sum_{k \geq 1} a_k z^k\Big|^2 \leq \Big(\sum_{k \geq 1} k|a_k|^2\Big)\Big(\sum_{k \geq 1} \frac{|z|^{2k}}{k}\Big) = \mathcal{D}(f) \log\Big(\frac{1}{1 - |z|^2}\Big). \quad \square$$

This result has some interesting consequences. The first is a growth estimate.

Corollary 1.2.2 *If $f \in \mathcal{D}$, then*

$$f(z) = o\Big(\Big(\log \frac{1}{1 - |z|^2}\Big)^{1/2}\Big) \qquad (|z| \to 1^-).$$

Proof Let $f(z) := \sum_{k \geq 0} a_k z^k$ and $s_n(z) := \sum_{k=0}^n a_k z^k$. Clearly s_n is bounded on $\mathbb{D}$. Hence, applying Theorem 1.2.1 to $f - s_n$, we obtain

$$\limsup_{|z| \to 1^-} |f(z)|\Big(\log \frac{1}{1 - |z|^2}\Big)^{-1/2} \leq \mathcal{D}(f - s_n)^{1/2}.$$

The result follows upon letting $n \to \infty$. $\square$

One might ask whether it is possible to go further, and prove that f must be bounded on $\mathbb{D}$. This is not the case. For example, consider the function

$$f(z) := \sum_{k \geq 2} \frac{z^k}{k \log k} \qquad (z \in \mathbb{D}).$$

Then

$$\mathcal{D}(f) = \sum_{k \geq 2} k\Big(\frac{1}{k \log k}\Big)^2 = \sum_{k \geq 2} \frac{1}{k(\log k)^2} < \infty,$$

so $f \in \mathcal{D}$ but, on the other hand,

$$\liminf_{r \to 1^-} f(r) \geq \sum_{k \geq 2} \frac{1}{k \log k} = \infty,$$

so f is unbounded on $\mathbb{D}$. (In fact, much more is true: see Exercise 1.2.2 below.)

A second consequence of Theorem 1.2.1 is that, for each $w \in \mathbb{D}$, the evaluation map $f \mapsto f(w)$ is a continuous linear functional on $\mathcal{D}$. By the Riesz representation theorem for Hilbert spaces, this functional is given by taking the inner product with some function $k_w \in \mathcal{D}$. We now give a direct proof of this, at the same time identifying k_w.

Theorem 1.2.3 *For $w \in \mathbb{D} \setminus \{0\}$, define*

$$k_w(z) := \frac{1}{z\overline{w}} \log\Big(\frac{1}{1 - z\overline{w}}\Big) \qquad (z \in \mathbb{D}),$$

and set $k_0 \equiv 1$. Then $k_w \in \mathcal{D}$ and

$$f(w) = \langle f, k_w \rangle_{\mathcal{D}} \qquad (w \in \mathbb{D}).$$

Proof Fix $w \in \mathbb{D}$. With k_w defined as in the statement of the theorem, we have $k_w(z) = \sum_{j\geq 0}(\overline{w}^j/(j+1))z^j$, the sum converging in the norm of $\mathcal{D}$. Therefore $k_w \in \mathcal{D}$. Further, if $f \in \mathcal{D}$, say $f(z) = \sum_{j\geq 0} a_j z^j$, then

$$\langle f, k_w \rangle_{\mathcal{D}} = \sum_{j\geq 0}(j+1)a_j \frac{w^j}{j+1} = \sum_{j\geq 0} a_j w^j = f(w). \qquad \square$$

The functions k_w are called *reproducing kernels*. They depend not only on the space $\mathcal{D}$ but also on our choice of inner product $\langle \cdot, \cdot \rangle_{\mathcal{D}}$. Later, we shall see that they play an important role in the study of zero sets (see §4.2) and Pick interpolation (see §5.3).

Exercises 1.2

1. Show that, for each $w \in \mathbb{D} \setminus \{0\}$,

$$\|k_w\|_{\mathcal{D}}^2 = \frac{1}{|w|^2} \log\Big(\frac{1}{1 - |w|^2}\Big).$$

2. Let $\epsilon : \mathbb{D} \to (0, \infty)$ be a function such that $\liminf_{|z|\to 1^-} \epsilon(z) = 0$. Show that there exists $f \in \mathcal{D}$ such that

$$f(z) \neq O\Big(\epsilon(z)\Big(\log \frac{1}{1 - |z|^2}\Big)^{1/2}\Big) \qquad (|z| \to 1).$$

[Hint: Let (w_n) be a sequence in $\mathbb{D}$ such that $|w_n| \to 1$ and $\epsilon(w_n) \to 0$. Use the result of Exercise 1 to show that the sequence

$$g_n := \epsilon(w_n)^{-1}\Big(\log \frac{1}{1 - |w_n|^2}\Big)^{-1/2} k_{w_n} \qquad (n \geq 1)$$

is unbounded in $\mathcal{D}$. By the Banach–Steinhaus theorem, there exists $f \in \mathcal{D}$ such that $\sup_n |\langle f, g_n \rangle_{\mathcal{D}}| = \infty$.]

1.3 Multiplication

Given $f, g \in \mathcal{D}$, we can form their pointwise product fg. It is natural to ask whether $fg \in \mathcal{D}$. The answer turns out to be negative.

Theorem 1.3.1 *The Dirichlet space is not an algebra.*

Proof We construct an explicit function f such that $f \in \mathcal{D}$ but $f^2 \notin \mathcal{D}$. Set

$$f(z) := \sum_{k\geq 2} \frac{z^k}{k(\log k)^{3/4}} \qquad (z \in \mathbb{D}).$$

By Theorem 1.1.2, we have

$$\mathcal{D}(f) = \sum_{k\geq 2} \frac{k}{(k(\log k)^{3/4})^2} = \sum_{k\geq 2} \frac{1}{k(\log k)^{3/2}} < \infty,$$

so $f \in \mathcal{D}$. On the other hand, $f(z)^2 = \sum_{k\geq 4} a_k z^k$ where

$$\begin{aligned} a_k &= \sum_{j=2}^{k-2} \frac{1}{j(\log j)^{3/4}(k-j)(\log(k-j))^{3/4}} \\ &\geq \frac{1}{k(\log k)^{3/4}} \sum_{j=2}^{k-2} \frac{1}{j(\log j)^{3/4}} \\ &\geq \frac{C}{k(\log k)^{1/2}}, \end{aligned}$$

for some constant $C > 0$ (see Exercise 1.3.1). Hence, by Theorem 1.1.2 again,

$$\mathcal{D}(f^2) \geq C^2 \sum_{k\geq 4} \frac{1}{k \log k} = \infty,$$

and so $f^2 \notin \mathcal{D}$. □

We do however have a positive result. In what follows, H^∞ denotes the algebra of bounded holomorphic functions on the unit disk with the norm $\|f\|_{H^\infty} := \sup_{z\in\mathbb{D}} |f(z)|$.

Theorem 1.3.2 *The space $\mathcal{D} \cap H^\infty$ is a Banach algebra with respect to the norm*

$$\|f\|_{\mathcal{D}\cap H^\infty} := \|f\|_{H^\infty} + \mathcal{D}(f)^{1/2}. \tag{1.2}$$

Proof Let $f, g \in \mathcal{D} \cap H^\infty$. Using Minkowski's inequality, we have

$$\begin{aligned}\mathcal{D}(fg)^{1/2} &= \Big(\frac{1}{\pi}\int_{\mathbb{D}} |f'g + fg'|^2\,dA\Big)^{1/2} \\ &\le \Big(\frac{1}{\pi}\int_{\mathbb{D}} |f'g|^2\,dA\Big)^{1/2} + \Big(\frac{1}{\pi}\int_{\mathbb{D}} |fg'|^2\,dA\Big)^{1/2} \\ &\le \mathcal{D}(f)^{1/2}\|g\|_{H^\infty} + \|f\|_{H^\infty}\mathcal{D}(g)^{1/2}.\end{aligned}$$

From this, it follows that $\mathcal{D}\cap H^\infty$ is an algebra, and that (1.2) defines an algebra norm on it. Finally, as both $(\mathcal{D}, \|\cdot\|_{\mathcal{D}})$ and $(H^\infty, \|\cdot\|_{H^\infty})$ are complete spaces, it is easy to check that $(\mathcal{D}\cap H^\infty, \|\cdot\|_{\mathcal{D}\cap H^\infty})$ is complete. □

In the light of this theorem, it is tempting to believe that $\mathcal{D}$ is stable under multiplication by elements of $\mathcal{D} \cap H^\infty$. We shall see later that this is actually false. The functions that do have this important property are called multipliers, and we shall return to study them in detail in Chapter 5.

Exercises 1.3

1. Let $\alpha \in (0, 1)$. By comparing the sum with an integral, show that

$$\sum_{j=2}^{k} \frac{1}{j(\log j)^\alpha} = \frac{(\log k)^{1-\alpha}}{1-\alpha} + O(1) \quad \text{as } k \to \infty.$$

1.4 Composition

This short section is based on the following conformal invariance property.

Theorem 1.4.1 *Let D_1, D_2 be domains, let $\phi : D_1 \to D_2$ be a conformal mapping and let $f : D_2 \to \mathbb{C}$ be a holomorphic function. Then*

$$\int_{D_1} |(f\circ\phi)'(z)|^2\,dA(z) = \int_{D_2} |f'(w)|^2\,dA(w).$$

Proof Making the substitution $w = \phi(z)$, we have $dA(w) = |\phi'(z)|^2\,dA(z)$, whence

$$\int_{D_2} |f'(w)|^2\,dA(w) = \int_{D_1} |f'(\phi(z))|^2\,|\phi'(z)|^2\,dA(z) = \int_{D_1} |(f\circ\phi)'(z)|^2\,dA(z). \quad \square$$

In particular, taking $D_1 = \mathbb{D}$ and $f(z) = z$, we obtain the following interpretation of the Dirichlet integral.

Corollary 1.4.2 *Let $\phi : \mathbb{D} \to \mathbb{C}$ be an injective holomorphic map. Then $\mathcal{D}(\phi)$ equals $1/\pi$ times the area of $\phi(\mathbb{D})$.*

Perhaps the most important case of Theorem 1.4.1 is when $D_1 = D_2 = \mathbb{D}$, in other words, when ϕ is a conformal automorphism of $\mathbb{D}$. The automorphisms of $\mathbb{D}$ are precisely the Möbius transformations of the form

$$\phi(z) = e^{i\theta}\frac{a-z}{1-\overline{a}z} \qquad (a \in \mathbb{D},\ |e^{i\theta}| = 1).$$

We write $\mathrm{Aut}(\mathbb{D})$ for this family of functions.

Corollary 1.4.3 *If $f \in \mathrm{Hol}(\mathbb{D})$ and $\phi \in \mathrm{Aut}(\mathbb{D})$, then $\mathcal{D}(f \circ \phi) = \mathcal{D}(f)$. Consequently, if $f \in \mathcal{D}$, then also $f \circ \phi \in \mathcal{D}$.*

The same argument shows that, if $f \in \mathcal{D}$ and $\phi : \mathbb{D} \to \mathbb{D}$ is any injective holomorphic function, then $f \circ \phi \in \mathcal{D}$. To what extent can the hypothesis 'injective' be weakened? We shall return to this question in Chapter 6. In the same chapter we shall also see that, quite remarkably, the property of Möbius invariance described in Corollary 1.4.3 essentially characterizes the Dirichlet space.

Exercises 1.4

1. Use Corollary 1.4.2 to give another example of an unbounded function in $\mathcal{D}$.

1.5 Douglas' formula

Let $\mathbb{T}$ denote the unit circle. Given $f \in \mathrm{Hol}(\mathbb{D})$ and $\zeta \in \mathbb{T}$, we write $f^*(\zeta) := \lim_{r \to 1^-} f(r\zeta)$, whenever this radial limit exists. If $f \in H^2$, then $f^*(\zeta)$ exists a.e. on $\mathbb{T}$. Moreover, $f^* \in L^2(\mathbb{T})$ and f is the Poisson integral of f^*. The correspondence $f \leftrightarrow f^*$ allows us to view the Hardy space as a space of functions on the unit circle, and this turns out to be vital for many applications.

There is a formula for $\mathcal{D}(f)$ expressed purely in terms of f^*. It is due to Douglas [39].

Theorem 1.5.1 (Douglas' formula) *Let $f \in H^2$. Then*

$$\mathcal{D}(f) = \frac{1}{4\pi^2}\int_{\mathbb{T}}\int_{\mathbb{T}}\left|\frac{f^*(\lambda) - f^*(\zeta)}{\lambda - \zeta}\right|^2 |d\lambda|\,|d\zeta|.$$

We shall deduce this formula from a lemma about L^2-Fourier series. Given a function $\phi \in L^2(\mathbb{T})$, we write $\widehat{\phi}(k)$ for its k-th Fourier coefficient, namely

$$\widehat{\phi}(k) := \frac{1}{2\pi}\int_0^{2\pi} \phi(e^{it})e^{-ikt}\,dt \qquad (k \in \mathbb{Z}).$$

In this notation, Parseval's formula becomes

$$\frac{1}{2\pi}\int_0^{2\pi}|\phi(e^{it})|^2\,dt=\sum_{k\in\mathbb{Z}}|\widehat{\phi}(k)|^2.$$

Lemma 1.5.2 *Let $\phi\in L^2(\mathbb{T})$. Then*

$$\frac{1}{4\pi^2}\int_{\mathbb{T}}\int_{\mathbb{T}}\Bigl|\frac{\phi(\lambda)-\phi(\zeta)}{\lambda-\zeta}\Bigr|^2|d\lambda|\,|d\zeta|=\sum_{k\in\mathbb{Z}}|k|\,|\widehat{\phi}(k)|^2.$$

Proof After the change of variables $\lambda=e^{i(s+t)}$, $\zeta=e^{it}$, the double integral becomes

$$\frac{1}{4\pi^2}\int_0^{2\pi}\int_0^{2\pi}\Bigl|\frac{\phi(e^{i(s+t)})-\phi(e^{it})}{e^{is}-1}\Bigr|^2\,dt\,ds.$$

Parseval's formula, applied to the function $\zeta\mapsto\phi(e^{is}\zeta)-\phi(\zeta)$, gives

$$\frac{1}{2\pi}\int_0^{2\pi}|\phi(e^{i(s+t)})-\phi(e^{it})|^2\,dt=\sum_{k\in\mathbb{Z}}|\widehat{\phi}(k)|^2|e^{iks}-1|^2.$$

Hence the double integral equals

$$\frac{1}{2\pi}\int_0^{2\pi}\sum_{k\in\mathbb{Z}}|\widehat{\phi}(k)|^2\frac{|e^{iks}-1|^2}{|e^{is}-1|^2}\,ds.$$

Finally, we remark that, for each integer $k\neq 0$,

$$\frac{1}{2\pi}\int_0^{2\pi}\Bigl|\frac{e^{iks}-1}{e^{is}-1}\Bigr|^2\,ds=\frac{1}{2\pi}\int_0^{2\pi}|1+e^{is}+\cdots+e^{i(|k|-1)s}|^2\,ds=|k|,$$

the last equality again from Parseval's formula. The result follows. □

Proof of Theorem 1.5.1 We apply Lemma 1.5.2 with $\phi=f^*$. Observe that, writing $f(z)=\sum_{k\geq 0}a_kz^k$, we have $\widehat{f^*}(k)=a_k$ if $k\geq 0$ and $\widehat{f^*}(k)=0$ if $k<0$. Hence

$$\frac{1}{4\pi^2}\int_{\mathbb{T}}\int_{\mathbb{T}}\Bigl|\frac{f^*(\lambda)-f^*(\zeta)}{\lambda-\zeta}\Bigr|^2|d\lambda|\,|d\zeta|=\sum_{k\geq 0}k|a_k|^2.$$

By Theorem 1.1.2, this last expression equals $\mathcal{D}(f)$. □

As mentioned at the beginning of the section, if $f\in H^2$, then the radial limit f^* exists a.e. Indeed, for almost every $\zeta\in\mathbb{T}$, we have $f(z)\to f^*(\zeta)$ as $z\to\zeta$ in each non-tangential approach region $|z-\zeta|<\kappa(1-|z|)$. For functions in $\mathcal{D}$, the same is true even for certain tangential approach regions. We shall prove this as an application of Douglas' formula.

Theorem 1.5.3 *Let $f\in\mathcal{D}$. Then, for a.e. $\zeta\in\mathbb{T}$, we have $f(z)\to f^*(\zeta)$ as $z\to\zeta$ in each oricyclic approach region $|z-\zeta|<\kappa(1-|z|)^{1/2}$.*

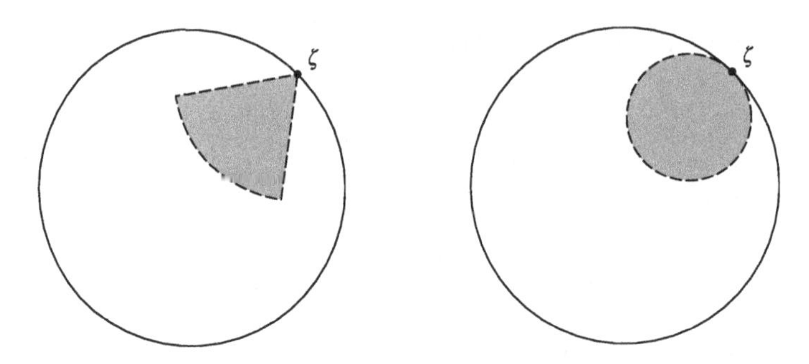

Figure 1.1 Non-tangential and oricyclic approach regions

Lemma 1.5.4 *If $g \in H^2$, then* $\lim_{|z|\to 1}(1-|z|^2)|g(z)|^2 = 0$.

Proof Write $g(z) = \sum_{k\geq 0} a_k z^k$. By the Cauchy–Schwarz inequality,

$$|g(z)|^2 \leq \sum_{k\geq 0} |a_k|^2 \sum_{l\geq 0} |z|^{2l} = \|g\|_{H^2}^2 (1-|z|^2)^{-1} \qquad (z \in \mathbb{D}).$$

Hence $\limsup_{|z|\to 1}(1-|z|^2)|g(z)|^2 \leq \|g\|_{H^2}^2$. Replacing g by $g - \sum_{k=0}^n a_k z^k$, and then letting $n \to \infty$, we deduce that in fact $\lim_{|z|\to 1}(1-|z|^2)|g(z)|^2 = 0$. □

Proof of Theorem 1.5.3 Let $f \in \mathcal{D}$. By Douglas' formula, we have

$$\int_{\mathbb{T}}\int_{\mathbb{T}} \left|\frac{f^*(\lambda) - f^*(\zeta)}{\lambda - \zeta}\right|^2 |d\lambda|\,|d\zeta| < \infty.$$

Hence, for almost every $\zeta \in \mathbb{T}$, the radial limit $f^*(\zeta)$ exists and satisfies

$$\int_{\mathbb{T}} \left|\frac{f^*(\lambda) - f^*(\zeta)}{\lambda - \zeta}\right|^2 |d\lambda| < \infty.$$

Fix such a ζ, and define

$$g(z) := \frac{f(z) - f^*(\zeta)}{z - \zeta} \qquad (z \in \mathbb{D}).$$

Then $g \in \mathrm{Hol}(\mathbb{D})$, the radial limit g^* exists a.e., and $g^* \in L^2(\mathbb{T})$. Since the denominator of g is an outer function, Smirnov's maximum principle implies that $g \in H^2$ (see Theorem A.3.8 in Appendix A). Therefore Lemma 1.5.4 applies, and $|g(z)|^2 = o((1-|z|^2)^{-1})$ as $|z| \to 1$. It follows that

$$|f(z) - f^*(\zeta)|^2 = |z-\zeta|^2 |g(z)|^2 = o\Big(\frac{|z-\zeta|^2}{1-|z|^2}\Big) \qquad (|z| \to 1).$$

Hence $f(z) \to f^*(\zeta)$ as $z \to \zeta$ in each approach region $|z-\zeta| < \kappa(1-|z|)^{1/2}$. □

There is a lot more to be said about boundary values of functions in $\mathcal{D}$. We shall treat this topic in some detail in Chapter 3, where, in particular, we shall see that the tangential approach region in Theorem 1.5.3 can be widened considerably further.

1.6 Weighted Dirichlet spaces

In this section we briefly consider the following generalization of the notions of Dirichlet integral and Dirichlet space.

Definition 1.6.1 Let $w : \mathbb{D} \to [0, \infty)$ be a measurable function such that $\int_{\mathbb{D}} w(z)\, dA(z) < \infty$. Given $f \in \mathrm{Hol}(\mathbb{D})$, we define its *weighted Dirichlet integral* by

$$\mathcal{D}_w(f) := \frac{1}{\pi} \int_{\mathbb{D}} |f'(z)|^2 w(z)\, dA(z).$$

The corresponding *weighted Dirichlet space* $\mathcal{D}_w$ is the vector space of those $f \in \mathrm{Hol}(\mathbb{D})$ such that $\mathcal{D}_w(f) < \infty$.

Obviously, taking $w \equiv 1$, we recover the standard Dirichlet space $\mathcal{D}$. The following theorem yields another familiar example.

Theorem 1.6.2 *Let* $w(z) = \log(1/|z|^2)$. *Then* $\mathcal{D}_w = H^2$ *and*

$$\|f\|_{H^2}^2 = |f(0)|^2 + \mathcal{D}_w(f) \qquad (f \in H^2).$$

Proof This is a computation like that in Theorem 1.1.2. Let $f(z) = \sum_{k \geq 0} a_k z^k$. Then

$$\begin{aligned}\mathcal{D}_w(f) &= \frac{1}{\pi} \int_0^1 \int_0^{2\pi} \Big|\sum_{k \geq 1} k a_k r^{k-1} e^{i(k-1)\theta}\Big|^2 \log\Big(\frac{1}{r^2}\Big)\, d\theta\, r\, dr \\ &= 2 \int_0^1 \sum_{k \geq 1} k^2 |a_k|^2 r^{2k-2} \log(1/r^2) r\, dr = \sum_{k \geq 1} |a_k|^2.\end{aligned}$$

The result follows. □

Theorem 1.6.3 *Suppose that* $\liminf_{|z| \to 1^-} w(z)/(1 - |z|) > 0$. *Then* $\mathcal{D}_w \subset H^2$, *and* $\mathcal{D}_w$ *is a Hilbert space with respect to the norm* $\|\cdot\|_{\mathcal{D}_w}$ *given by*

$$\|f\|_{\mathcal{D}_w}^2 := \|f\|_{H^2}^2 + \mathcal{D}_w(f) \qquad (f \in \mathcal{D}_w). \tag{1.3}$$

Proof By hypothesis, there exist $c > 0$ and $r < 1$ such that $w(z) \geq c \log(1/|z|^2)$

for all z with $r < |z| < 1$. A calculation similar to that in the previous theorem shows that, if $f(z) = \sum_{k\geq 0} a_k z^k$, then

$$\mathcal{D}_w(f) \geq \frac{1}{\pi}\int_{r<|z|<1} |f'(z)|^2 c \log\Bigl(\frac{1}{|z|^2}\Bigr)\,dA(z) = c\sum_{k\geq 0}\Bigl(1 - r^{2k} - 2kr^{2k}\log\Bigl(\frac{1}{r}\Bigr)\Bigr)|a_k|^2.$$

It follows that $\mathcal{D}_w \subset H^2$.

It remains to prove that $(\mathcal{D}_w, \|\cdot\|_{\mathcal{D}_w})$ is complete. Let (f_n) be a Cauchy sequence with respect to $\|\cdot\|_{\mathcal{D}_w}$. Then it is Cauchy with respect to $\|\cdot\|_{H^2}$, so there exists $f \in H^2$ such that $f_n \to f$ in H^2. In particular, f is holomorphic in $\mathbb{D}$, and $f_n \to f$ locally uniformly in $\mathbb{D}$. Hence also $f_n' \to f'$ locally uniformly in $\mathbb{D}$. By Fatou's lemma, we have $\mathcal{D}_w(f) \leq \liminf_{n\to\infty} \mathcal{D}_w(f_n) < \infty$, so $f \in \mathcal{D}_w$. By Fatou's lemma again, $\|f - f_n\|_{\mathcal{D}_w} \leq \liminf_{m\to\infty} \|f_m - f_n\|_{\mathcal{D}_w}$, and from this it follows easily that $\|f - f_n\|_{\mathcal{D}_w} \to 0$ as $n \to \infty$. □

We now turn to study multiplication in $\mathcal{D}_w$.

Theorem 1.6.4 *$\mathcal{D}_w \cap H^\infty$ is a Banach algebra with respect to the norm*

$$\|f\|_{\mathcal{D}_w \cap H^\infty} := \|f\|_{H^\infty} + \mathcal{D}_w(f)^{1/2}.$$

Proof This virtually identical to the proof of Theorem 1.3.2. □

Corollary 1.6.5 *If $\mathcal{D}_w \subset H^\infty$, then $\mathcal{D}_w$ is an algebra.* □

A partial converse to this result is given in Exercise 1.6.6.

Theorem 1.6.6 *$\mathcal{D}_w \subset H^\infty$ (and consequently $\mathcal{D}_w$ is an algebra) if*

$$\sup_{\zeta\in\mathbb{T}} \int_{\mathbb{D}} \frac{dA(z)}{w(z)|\zeta - z|^2} < \infty.$$

In the course of the proof, we shall need the following elementary inequality. The same inequality will recur several times in the book.

Lemma 1.6.7 *Let $0 < r < 1$. Then*

$$\frac{|1 - z|}{|1 - rz|} \leq \frac{2}{1 + r} \qquad (z \in \overline{\mathbb{D}}).$$

Proof The Möbius transformation $z \mapsto (1 - z)/(1 - rz)$ maps the unit circle onto a circle symmetric with respect to the real axis and passing through 0 and $2/(1 + r)$. It is therefore the circle with diameter $[0, 2/(1 + r)]$. The result now follows by the maximum principle. □

Proof of Theorem 1.6.6 Let $f \in \mathcal{D}_w$ and let $\lambda \in \mathbb{D}$. Then, by the Cauchy–Schwarz inequality,

$$\int_{\mathbb{D}} \Big|\frac{f'(z)}{1-\bar{z}\lambda}\Big|\,dA(z) \le \Big(\int_{\mathbb{D}} |f'(z)|^2 w(z)\,dA(z)\Big)^{1/2}\Big(\int_{\mathbb{D}} \frac{dA(z)}{w(z)|1-\bar{z}\lambda|^2}\Big)^{1/2}$$
$$\le \Big(\int_{\mathbb{D}} |f'(z)|^2 w(z)\,dA(z)\Big)^{1/2}\Big(\int_{\mathbb{D}} \frac{4\,dA(z)}{w(z)|(\lambda/|\lambda|)-z|^2}\Big)^{1/2}.$$

Here, for the last inequality we used Lemma 1.6.7:

$$|(\lambda/|\lambda|) - z| = \big|1 - z\bar{\lambda}/|\lambda|\big| \le 2|1-\bar{z}\lambda| \qquad (\lambda, z \in \mathbb{D}).$$

Consequently, by our hypothesis on w,

$$h(\lambda) := \frac{1}{\pi}\int_{\mathbb{D}} \frac{f'(z)}{1-\bar{z}\lambda}\,dA(z) \qquad (\lambda \in \mathbb{D})$$

defines a bounded holomorphic function h on $\mathbb{D}$. Differentiating under the integral sign n times, we see that $h^{(n)}(0)/n! = f^{(n+1)}(0)/(n+1)!$ for all n. Therefore $h(\lambda) = (f(\lambda) - f(0))/\lambda$. We conclude that $f \in H^\infty$. □

We mention here two particularly important classes of weights. The first is the family of power weights $w(z) := (1-|z|^2)^\alpha$ $(-1 < \alpha \le 1)$. The corresponding weighted Dirichlet spaces are usually denoted $\mathcal{D}_\alpha$ (with $\mathcal{D}_\alpha(f)$ denoting the corresponding Dirichlet integral). For $0 \le \alpha \le 1$, they provide a natural scale of spaces linking the Dirichlet space $\mathcal{D}_0$ to the Hardy space $\mathcal{D}_1$ (see Exercise 1.6.1). For $-1 < \alpha < 0$ they are algebras, by Theorem 1.6.6. We shall not study them systematically, though they will feature in the Exercises.

The second class of weights is the family of harmonic weights. A positive harmonic function on $\mathbb{D}$ has the form $w = P\mu$, the Poisson integral of a positive finite measure μ on the unit circle. The corresponding harmonically weighted Dirichlet spaces are denoted $\mathcal{D}_\mu$. They turn out to be useful for a number of purposes, and we shall return to study them in detail in Chapter 7.

Exercises 1.6

1. Let $w(z) := 1 - |z|^2$. Show that $\mathcal{D}_w = H^2$ and that
$$\frac{1}{2}\|f\|_{H^2}^2 \le |f(0)|^2 + \mathcal{D}_w(f) \le \|f\|_{H^2}^2 \qquad (f \in H^2).$$
2. Use Theorem 1.6.3 to show that, if μ is a positive finite measure on $\mathbb{T}$, then $\mathcal{D}_\mu \subset H^2$.
3. Use Theorem 1.6.6 to show that, if $-1 < \alpha < 0$, then $\mathcal{D}_\alpha \subset H^\infty$.
4. (i) Let $\alpha \in (0,1)$. Show that $\int_0^1 t^{k-1}(1-t)^\alpha\,dt \asymp 1/k^{1+\alpha}$ for all $k \ge 1$, in the sense that the ratio of the two sides is bounded above and below by positive constants depending only on α.

(ii) Deduce that, if $f(z) = \sum_{k\geq 0} a_k z^k$, then

$$\mathcal{D}_\alpha(f) \asymp \sum_{k\geq 1} k^{1-\alpha}|a_k|^2.$$

5. Again, let $\alpha \in (0, 1)$. Show that, if $f \in H^2$, then

$$\mathcal{D}_\alpha(f) \asymp \int_{\mathbb{T}}\int_{\mathbb{T}} \frac{|f^*(\lambda) - f^*(\zeta)|^2}{|\lambda - \zeta|^{2-\alpha}}\,|d\lambda|\,|d\zeta|,$$

where the implied constants depend only on α. [Hint: Mimic the proof of Theorem 1.5.1.]

6. Suppose that $\mathcal{D}_w \subset H^2$ and that $\mathcal{D}_w$ is an algebra. Prove that $\mathcal{D}_w \subset H^\infty$ as follows. Let $f \in \mathcal{D}_w$.
   (i) Show that the map $g \mapsto fg : \mathcal{D}_w \to \mathcal{D}_w$ is continuous with respect to the norm $\|\cdot\|_{\mathcal{D}_w}$ defined by (1.3). [Hint: Closed graph theorem.]
   (ii) Deduce that there exists a constant C_f, such that $\|f^n\|_{\mathcal{D}_w} \leq C_f^n$ for all $n \geq 0$.
   (iii) Deduce that, for each $z \in \mathbb{D}$, there exists a constant C_z, such that $|f(z)|^n \leq C_z C_f^n$ for all $n \geq 0$.
   (iv) Deduce that $|f(z)| \leq C_f$ for all $z \in \mathbb{D}$.

Notes on Chapter 1

§1.1

As mentioned in the preface, the notion of Dirichlet space goes back at least to Beurling's doctoral thesis [15], which was published in 1933 (though parts of it were written a few years earlier). There are two recent survey articles treating the subject from different points of view, one by Ross [103], the other by Arcozzi, Rochberg, Sawyer and Wick [11].

§1.2

The notion of reproducing kernel makes sense for any Hilbert space of functions on a set X such that evaluation at each point of X is a non-zero continuous linear functional. For a detailed treatment of these ideas, we refer to the book of Agler and McCarthy [4].

§1.5

The formula in Theorem 1.5.1 (and its generalization to harmonic functions) played a key role in Douglas' solution to the Plateau problem [39], which earned him one of the two inaugural Fields medals in 1936.

2

Capacity

Central to the study of the Dirichlet space is the concept of logarithmic capacity. This assigns to subsets of the unit circle a notion of size which is closely linked to various aspects of functions in $\mathcal{D}$, notably boundary behavior, zeros, multipliers and cyclicity. It will thus be a recurring theme throughout the book.

In this chapter we present a brief but self-contained account of capacity. We do so in an abstract setting, replacing the unit circle by a general compact metric space, and the logarithmic kernel by an arbitrary decreasing function. This entails no extra work, and has the advantage that it covers not only logarithmic capacity but also certain other capacities such as Riesz capacities, which arise naturally in the context of the spaces $\mathcal{D}_\alpha$.

2.1 Potentials, energy and capacity

Throughout the chapter, we fix a compact metric space (X, d) and a continuous decreasing function $K : (0, \infty) \to [0, \infty)$. The function K is called a *kernel* (though it has nothing to do with reproducing kernels). We extend the definition of K to 0 by defining $K(0) := \lim_{t \to 0^+} K(t)$. It may well happen that $K(0) = \infty$, and in fact this is the case for most interesting kernels, though we do not insist upon it. However, in order to avoid trivialities, we do assume that $K \not\equiv 0$.

Definition 2.1.1 Let μ be a finite positive Borel measure on X. Its *potential* is the function $K\mu : X \to [0, \infty]$ defined by

$$K\mu(x) := \int_X K(d(x, y))\, d\mu(y) \qquad (x \in X).$$

The *energy* of μ is defined as

$$I_K(\mu) := \int K\mu\, d\mu = \iint K(d(x,y))\, d\mu(x)\, d\mu(y).$$

Clearly $I_K(\mu) \in [0, \infty]$. In fact $I_K(\mu) > 0$ (see Exercise 2.1.1). However, it is quite possible that $I_K(\mu) = \infty$.

Definition 2.1.2 Let F be a compact subset of X. We write $\mathcal{P}(F)$ for the set of Borel probability measures on F. The *capacity* of F is defined by

$$c_K(F) := 1/\inf\{I_K(\mu) : \mu \in \mathcal{P}(F)\}. \tag{2.1}$$

In particular $c_K(F) = 0$ if and only if there is no measure $\mu \in \mathcal{P}(F)$ with $I_K(\mu) < \infty$. The following result gives some basic properties of capacity as a set function.

Theorem 2.1.3 (i) $c_K(\emptyset) = 0$.
(ii) *If* $F_1 \subset F_2$, *then* $c_K(F_1) \le c_K(F_2)$.
(iii) $c_K(F_1 \cup \cdots \cup F_n) \le c_K(F_1) + \cdots + c_K(F_n)$.

Proof Parts (i) and (ii) are obvious. For (iii), it suffices to treat the case $n = 2$, and then use induction. Given $\mu \in \mathcal{P}(F_1 \cup F_2)$, we can write $\mu = t\mu_1 + (1-t)\mu_2$, where $\mu_j \in \mathcal{P}(F_j)$ for $j = 1, 2$ and $t \in [0, 1]$. (For example, take $t := \mu(F_1)$ and define $\mu_1(S) := \mu(S \cap F_1)/t$ and $\mu_2(S) := \mu(S \setminus F_1)/(1-t)$.) Then we have

$$\begin{aligned} I_K(\mu) &\ge t^2 I_K(\mu_1) + (1-t)^2 I_K(\mu_2) \\ &\ge \frac{t^2}{c_K(F_1)} + \frac{(1-t)^2}{c_K(F_2)} \\ &\ge \frac{1}{c_K(F_1) + c_K(F_2)}, \end{aligned}$$

where, for the last line, we used the inequality $t^2/a + (1-t)^2/b \ge 1/(a+b)$. Taking the infimum over all such μ, we obtain

$$\frac{1}{c_K(F_1 \cup F_2)} \ge \frac{1}{c_K(F_1) + c_K(F_2)},$$

and inverting this inequality gives the result. □

Next, here is a simple upper bound for capacity in terms of diameter.

Theorem 2.1.4 *If F is a compact subset of X, then* $c_K(F) \le 1/K(\mathrm{diam}(F))$.

Proof If $\mu \in \mathcal{P}(F)$, then

$$\begin{aligned} I_K(\mu) &= \iint K(d(x,y))\,d\mu(x)\,d\mu(y) \\ &\geq \iint K(\operatorname{diam}(F))\,d\mu(x)\,d\mu(y) = K(\operatorname{diam}(F)). \end{aligned}$$

It follows that $c_K(F) \leq 1/K(\operatorname{diam}(F))$. □

Corollary 2.1.5 *For every compact set F we have $c_K(F) < \infty$.*

Proof By assumption $K \not\equiv 0$, so there exists $d_0 > 0$ such that $K(d_0) > 0$. A compact set F can be covered by finitely many compact sets $F_1, \ldots, F_n$ of diameter at most d_0, so $c_K(F) \leq c_K(F_1) + \cdots + c_K(F_n) \leq n/K(d_0) < \infty$. □

The next result is sometimes expressed by saying that capacity is upper semicontinuous.

Theorem 2.1.6 *Let $(F_n)_{n\geq 1}$ be a decreasing sequence of compact subsets of X, and let $F := \cap_n F_n$. Then $c_K(F) = \lim_n c_K(F_n)$.*

The proof makes use of the notion of weak*-convergence in $\mathcal{P}(X)$. Recall that $\mu_n \to \mu$ weak* in $\mathcal{P}(X)$ if $\int f\,d\mu_n \to \int f\,d\mu$ for each continuous function f on X. Also, every sequence (μ_n) in $\mathcal{P}(X)$, contains a weak*-convergent subsequence.

Lemma 2.1.7 *Let (μ_n) be a sequence in $\mathcal{P}(X)$, and suppose that μ_n is weak*-convergent to $\mu \in \mathcal{P}(X)$. Then $\liminf_{n\to\infty} I_K(\mu_n) \geq I_K(\mu)$.*

Proof We first claim that, if $f : X \times X \to \mathbb{R}$ is a continuous function, then $\iint f\,d\mu_n\,d\mu_n \to \iint f\,d\mu\,d\mu$ as $n \to \infty$. Indeed, this is clearly true if f has the form $f(x,y) = g(x)h(y)$ and, using the Stone–Weierstrass theorem, a general continuous f may be uniformly approximated by finite sums of functions of this special form.

If $K(0) < \infty$, then we can apply the claim with $f(x,y) := K(d(x,y))$, and deduce that $I_K(\mu_n) \to I_K(\mu)$. For the general case, we consider $K_T(t) := \min\{K(t), T\}$. By what we have just proved, $I_{K_T}(\mu) = \lim_{n\to\infty} I_{K_T}(\mu_n)$. Clearly $I_{K_T}(\mu_n) \leq I_K(\mu_n)$ for all n, and so $I_{K_T}(\mu) \leq \liminf_{n\to\infty} I_K(\mu_n)$. Also, by the monotone convergence theorem $I_{K_T}(\mu) \to I_K(\mu)$ as $T \to \infty$. Hence $I_K(\mu) \leq \liminf_{n\to\infty} I_K(\mu_n)$, as required. □

Proof of Theorem 2.1.6 By Theorem 2.1.3 (ii) the sequence $c_K(F_n)$ decreases and $c_K(F_n) \geq c_K(F)$ for all n. We must show that $\lim_{n\to\infty} c_K(F_n) \leq c_K(F)$.

If $c_K(F_n) = 0$ for some n, then we are done. If not, then, for each n, pick $\mu_n \in \mathcal{P}(F_n)$ such that $I_K(\mu_n) < 1/c_K(F_n) + 1/n$. There exists a subsequence of the μ_n which is weak*-convergent to some $\mu \in \mathcal{P}(X)$. Relabeling, if necessary, we may as well suppose that this is the whole sequence. Note that μ is supported in F_n for each n, so $\mu \in \mathcal{P}(F)$ and $I_K(\mu) \geq 1/c_K(F)$. On the other hand, by Lemma 2.1.7, we have $I_K(\mu) \leq \liminf_{n\to\infty} I_K(\mu_n) \leq \lim_{n\to\infty} 1/c_K(F_n)$. It follows that $\lim_{n\to\infty} c_K(F_n) \leq c_K(F)$. □

Now we extend the definition of capacity to non-compact sets.

Definition 2.1.8 Let E be an arbitrary subset of X.

- The *inner capacity* of E is defined by

$$c_K(E) := \sup\{c_K(F) : F \subset E,\ F \text{ compact}\}.$$

- The *outer capacity* of E is defined by

$$c_K^*(E) := \inf\{c_K(U) : U \supset E,\ U \text{ open in } X\}.$$

Evidently $c_K(E) \leq c_K^*(E)$. A set E is called *capacitable* if $c_K(E) = c_K^*(E)$. It is clear that open sets are capacitable, and Theorem 2.1.6 shows that compact sets are capacitable too. In fact, for several of the most important kernels K, it can be shown that every Borel subset of X is capacitable (Choquet's theorem), but we do not need that result here.

The next theorem summarizes some basic properties of c_K^*.

Theorem 2.1.9 (i) $c_K^*(\emptyset) = 0$.
(ii) *If* $E_1 \subset E_2$, *then* $c_K^*(E_1) \leq c_K^*(E_2)$.
(iii) $c_K^*(\cup_{k\geq 1} E_k) \leq \sum_{k\geq 1} c_K^*(E_k)$.

In particular, the outer capacity of a set does not change if we adjoin to it a set of outer capacity zero. Also, a countable union of sets of outer capacity zero is still of outer capacity zero.

Proof Parts (i) and (ii) are obvious.

For part (iii), we consider first the case when each E_k is an open set U_k. Let F be a compact subset of $\cup_k U_k$. By compactness there exists $n \geq 1$ such that $F \subset U_1 \cup \cdots \cup U_n$. We can write $F = F_1 \cup \cdots \cup F_n$, where each F_k is a compact subset of U_k (see Exercise 2.1.2). Then, using Theorem 2.1.3 (iii), we have

$$c_K(F) \leq \sum_{k=1}^{n} c_K(F_k) \leq \sum_{k=1}^{n} c_K(U_k) \leq \sum_{k\geq 1} c_K(U_k).$$

As this holds for each such F, we obtain $c_K(\cup_k U_k) \leq \sum_k c_K(U_k)$, proving the result in this case.

For the general case, we may suppose that $c_K^*(E_k) < \infty$ for all k, otherwise there is nothing to prove. Let $\epsilon > 0$ and, for each k, let U_k be an open set such that $U_k \supset E_k$ and $c(U_k) < c^*(E_k) + \epsilon/2^k$. Then $\cup_k U_k$ is an open set containing $\cup_k E_k$ and, by what we have already proved,

$$c_K(\cup_k U_k) \le \sum_k c_K(U_k) \le \sum_k c_K^*(E_k) + \epsilon.$$

Thus $c_K^*(\cup_k E_k) \le \sum_k c_K^*(E_k) + \epsilon$. This holds for all $\epsilon > 0$, whence the result. □

Corollary 2.1.10 *If $K(0) = \infty$, then $c_K^*(E) = 0$ for every countable set $E \subset X$.*

Proof The hypothesis that $K(0) = \infty$ implies that singletons are of capacity zero. The general result follows by applying Theorem 2.1.9 (iii). □

Exercises 2.1

1. Let K be a kernel with $K \not\equiv 0$, and let μ be a finite positive measure on X. Prove that $K\mu(x) > 0$ for all $x \in \operatorname{supp}\mu$, and deduce that $I_K(\mu) > 0$.
2. Justify the following assertion, used in the proof of Theorem 2.1.9. If F is a compact set and $U_1, \dots, U_n$ are open sets such that $F \subset U_1 \cup \dots \cup U_n$, then we can write $F = F_1 \cup \dots \cup F_n$, where $F_1, \dots, F_n$ are compact sets such that $F_k \subset U_k$ for all k.

2.2 Equilibrium measures

In this section we consider the important class of measures for which the infimum is attained in (2.1).

Definition 2.2.1 Let F be a compact subset of X such that $c_K(F) > 0$. An *equilibrium measure* for F is a measure $\nu \in \mathcal{P}(F)$ such that $I_K(\mu) \ge I_K(\nu)$ for all $\mu \in \mathcal{P}(F)$.

Clearly, if ν is an equilibrium measure for F, then $c_K(F) = 1/I_K(\nu)$.

Theorem 2.2.2 *Let F be a compact subset of X with $c_K(F) > 0$. Then F has an equilibrium measure.*

Proof For each $n \ge 1$, choose $\mu_n \in \mathcal{P}(F)$ such that $I_K(\mu_n) < 1/c_K(F) + 1/n$. A subsequence of the μ_n converges weak* to some $\nu \in \mathcal{P}(F)$. By Lemma 2.1.7, this ν must satisfy $I_K(\nu) \le 1/c_K(F)$, so it is an equilibrium measure for F. □

For many kernels K, it turns out that the equilibrium measure is unique. We shall not pursue this here. We do however need the following version of a theorem of Frostman, sometimes called the fundamental theorem of potential theory, which details some special properties of the potential of an equilibrium measure. We write $\operatorname{supp}\mu$ for the (closed) support of a measure μ.

Theorem 2.2.3 *Let F be a compact subset of X such that $c_K(F) > 0$. Let ν be an equilibrium measure for F. Then:*

(i) $K\nu(x) \le I_K(\nu)$ *for all* $x \in \operatorname{supp}\nu$,
(ii) $K\nu(x) \ge I_K(\nu)$ *for all* $x \in F \setminus E$, *where* $c_K^*(E) = 0$.

Lemma 2.2.4 *If μ is a finite positive Borel measure on X, then its potential $K\mu$ is a lower semicontinuous function on X.*

Proof Let $x_n \to x_0$ in X. By Fatou's lemma, we have

$$\liminf_{n\to\infty} \int K(d(x_n, y))\, d\mu(y) \ge \int K(d(x_0, y))\, d\mu(y).$$

In other words $\liminf_{n\to\infty} K\mu(x_n) \ge K\mu(x_0)$, as required. □

Proof of Theorem 2.2.3 We first claim that, if $\mu \in \mathcal{P}(F)$ and $I_K(\mu) < \infty$, then $\int K\nu\, d\mu \ge I_K(\nu)$. In showing this, we may as well suppose that $\int K\nu\, d\mu < \infty$, otherwise there is nothing to prove. For each $t \in (0, 1)$ consider the measure $\mu_t := t\mu + (1 - t)\nu$. A simple calculation using the identity $\int K\mu\, d\nu = \int K\nu\, d\mu$ shows that

$$I_K(\mu_t) = I_K(\nu) + 2t\Big(\int K\nu\, d\mu - I_K(\nu)\Big) + t^2\Big(I_K(\mu) + I_K(\nu) - 2\int K\nu\, d\mu\Big).$$

Also, since $\mu_t \in \mathcal{P}(F)$, it must satisfy $I_K(\mu_t) \ge I_K(\nu)$ for all $t \in (0, 1)$. It follows that $\int K\nu\, d\mu \ge I_K(\nu)$, as claimed.

(i) Suppose, if possible, that $K\nu(x) > I_K(\nu)$ for some $x \in \operatorname{supp}\nu$. As $K\nu$ is lower semicontinuous, there exists a neighborhood N of x such that $K\nu > I_K(\nu)$ on N. As $x \in \operatorname{supp}\nu$, we must have $\nu(N) > 0$, and hence

$$\int_N K\nu\, d\nu > I_K(\nu)\nu(N).$$

Also, applying the claim at the beginning of the proof to the measure $\mu(S) := \nu(S \setminus N)/\nu(X \setminus N)$, we have

$$\int_{X\setminus N} K\nu\, d\nu \ge I_K(\nu)\nu(X \setminus N).$$

Adding together the two inequalities, we obtain

$$\int K\nu\, d\nu > I_K(\nu),$$

which is clearly a contradiction. Hence $K\nu(x) \le I_K(\nu)$ for all $x \in \operatorname{supp}\nu$.

(ii) For each $n \ge 1$, set

$$E_n := \{x \in F : K\nu(x) \le I_K(\nu) - 1/n\}.$$

As $K\nu$ is lower semicontinuous, E_n is compact. Suppose, if possible, that $c_K(E_n) > 0$ for some n. Then E_n possesses an equilibrium measure ν_n. By the claim at the beginning of the proof, we have

$$\int K\nu \, d\nu_n \geq I_K(\nu).$$

On the other hand, since $K\nu \leq I_K(\nu) - 1/n$ on E_n, we have

$$\int K\nu \, d\nu_n \leq I_K(\nu) - 1/n.$$

This is clearly a contradiction. Thus, in fact, $c_K(E_n) = 0$ for all n. Setting $E := \cup_{n\geq 1} E_n$, we therefore have $c^*(E) = 0$ and $K\nu(x) \geq I_K(\nu)$ for all $x \in F \setminus E$. □

2.3 Cantor sets

We have seen in Corollary 2.1.10 that, provided $K(0) = \infty$, all countable sets are of capacity zero. What about uncountable sets? In this section we consider a family of Cantor-type sets that are compact and uncountable. Our aim is to determine which of these sets have capacity zero. To this end, we need a general estimate for capacity, which is of interest in its own right.

Definition 2.3.1 Given a compact subset F of X and $t > 0$, we write $N_F(t)$ for the *t-covering number* of F, namely the smallest number of sets of diameter at most t needed to cover F.

Theorem 2.3.2 *Let F be a compact subset of X. Then*

$$\frac{1}{c_K(F)} \geq \int_0^\infty \frac{|dK(t)|}{N_F(t)}.$$

In particular, if the integral diverges, then $c_K(F) = 0$.

For the proof we need the following simple lemma. In what follows, we write $B(x,t) := \{y \in X : d(x,y) \leq t\}$ and $K(\infty) := \lim_{t\to\infty} K(t)$.

Lemma 2.3.3 *Let μ be a probability measure on X. Then*

$$I_K(\mu) = \int_0^\infty \Big(\int \mu(B(x,t))\, d\mu(x) \Big) |dK(t)| + K(\infty). \tag{2.2}$$

Proof Using Fubini's theorem, we have

$$\begin{aligned}
I_K(\mu) - K(\infty) &= \iint (K(d(x,y)) - K(\infty))\, d\mu(x)\, d\mu(y) \\
&= \iint \Big(\int_{t=d(x,y)}^{\infty} |dK(t)| \Big)\, d\mu(x)\, d\mu(y) \\
&= \int_0^{\infty} \Big(\iint_{d(x,y)\le t} d\mu(y)\, d\mu(x) \Big) |dK(t)| \\
&= \int_0^{\infty} \Big(\int \mu(B(x,t))\, d\mu(x) \Big) |dK(t)|. \qquad \square
\end{aligned}$$

Proof of Theorem 2.3.2 Let μ be a probability measure on F. Let $t > 0$, let $N = N_F(t)$, and let $A_1, \dots, A_N$ be a cover of F by closed sets of diameter $\le t$. Set $B_1 := F \cap A_1$ and $B_j := (F \cap A_j) \setminus (A_1 \cup \cdots \cup A_{j-1})$ $(j \ge 2)$. Then $B_1, \dots, B_N$ is a Borel partition of F. Also, if $x \in B_j$ then $B_j \subset B(x,t)$. Consequently,

$$\begin{aligned}
\int \mu(B(x,t))\, d\mu(x) &= \sum_{j=1}^{N} \int_{B_j} \mu(B(x,t))\, d\mu(x) \\
&\ge \sum_{j=1}^{N} \int_{B_j} \mu(B_j)\, d\mu(x) = \sum_{j=1}^{N} \mu(B_j)^2.
\end{aligned}$$

Now $1 = \mu(F)^2 = \big(\sum_{j=1}^N \mu(B_j)\big)^2 \le N \sum_{j=1}^N \mu(B_j)^2$, by the Cauchy–Schwarz inequality. Hence

$$\int \mu(B(x,t))\, d\mu(x) \ge \frac{1}{N} = \frac{1}{N_F(t)}.$$

Substituting this information into (2.2), we deduce that

$$I_K(\mu) \ge \int_0^{\infty} \frac{|dK(t)|}{N_F(t)}.$$

As this holds for all probability measures μ on F, the result follows. $\square$

We now define the Cantor-type sets that we are going to use.

Definition 2.3.4 Let $\{E_n^j : 1 \le j \le 2^n,\ n \ge 0\}$ be non-empty, compact subsets of X such that, for each $n \ge 1$:

- the sets E_n^j $(j = 1, \dots, 2^n)$ are pairwise disjoint, and
- each set E_{n-1}^i contains precisely two of the sets E_n^j.

Then the *Cantor set* corresponding to the data (E_n^j) is $E := \cap_{n \ge 0} \cup_{j=1}^{2^n} E_n^j$.

Clearly E is an uncountable compact set. We do not assume that it is totally disconnected, though this is often the case.

Theorem 2.3.5 *Let E be the Cantor set corresponding to (E_n^j). Then*

$$\sum_{n\geq 0} \frac{K(d_n)}{2^{n+1}} \leq \frac{1}{c_K(E)} \leq \sum_{n\geq 0} \frac{K(e_n)}{2^n}, \tag{2.3}$$

where

$$d_n := \max\{\operatorname{diam} E_n^j : 1 \leq j \leq 2^n\}$$
$$e_n := \min\{\operatorname{dist}(E_{n+1}^j, E_{n+1}^k) : E_{n+1}^j, E_{n+1}^k \subset \textit{the same } E_n^i,\ j \neq k\}.$$

In particular, if the left-hand series in (2.3) diverges, then $c_K(E) = 0$, and if the right-hand series converges, then $c_K(E) > 0$.

Proof For convenience, we suppose that $K(\infty) = 0$. This involves no loss of generality since, if (2.3) holds for such a kernel K, then it also holds for $K + a$ for each positive constant a.

Clearly, if $t \geq d_n$ then $N_E(t) \leq 2^n$. Hence, by Theorem 2.3.2, we have

$$\frac{1}{c_K(E)} \geq \int_0^\infty \frac{|dK(t)|}{N_E(t)} \geq K(d_0) + \sum_{n\geq 1} \frac{K(d_n) - K(d_{n-1})}{2^n} = \sum_{n\geq 0} \frac{K(d_n)}{2^{n+1}}.$$

This proves the left-hand inequality in (2.3).

For the right-hand inequality, we consider a measure μ on E satisfying $\mu(E_n^j) = 1/2^n$ for all j, n (see Exercise 2.3.1). By Lemma 2.3.3,

$$I_K(\mu) \leq K(\epsilon_1) + \sum_{n\geq 1} \int_{\epsilon_{n+1}}^{\epsilon_n} \int \mu(B(x,t))\, d\mu(x)\, |dK(t)|,$$

where $(\epsilon_n)_{n\geq 1}$ is any decreasing sequence which tends to zero. We apply this with $\epsilon_n := \min\{e_0, \ldots, e_{n-1}\}$. If $x \in E$ and $t < \epsilon_n$, then $B(x,t)$ meets E_n^j for only one j, so $\mu(B(x,t)) \leq 1/2^n$. Consequently

$$I_K(\mu) \leq K(\epsilon_1) + \sum_{n\geq 1} \frac{K(\epsilon_{n+1}) - K(\epsilon_n)}{2^n} \leq \sum_{n\geq 0} \frac{K(e_n)}{2^n},$$

the final inequality because $K(\epsilon_{n+1}) - K(\epsilon_n)$ is either zero or $K(e_n) - K(\epsilon_n)$. By definition, $1/c_K(E) \leq I_K(\mu)$. This gives the right-hand side of (2.3). □

Exercises 2.3

1. Let E be the Cantor set as constructed above. Prove that there exists a Borel probability measure μ on E such that $\mu(E_n^j) = 1/2^n$ for all j, n. Under what circumstances is μ unique?

2.4 Logarithmic capacity

Though the notion of capacity has been developed in some generality, we shall be interested mostly in the following special case. In what follows, we write $\log^+(x) := \max\{\log x, 0\}$.

Definition 2.4.1 Let $X = \mathbb{T}$ with the chordal metric $d(z,w) := |z-w|$, and let $K(t) := \log^+(2/t)$. The corresponding capacity is called *logarithmic capacity*. We shall denote it simply by $c(\cdot)$ and the associated outer capacity by $c^*(\cdot)$.

The choice of the constant 2 in $\log^+(2/t)$ is largely for convenience, since $\mathbb{T}$ has diameter 2. It could be replaced by any other constant $A \geq 2$, and the resulting capacity c_A would be equivalent in the sense that $1/c_A(F) - 1/c(F) = \log(A/2)$ for all F.

Another possibility, often seen in the literature, is to take $K(t) = \log(1/t)$. This gives a non-positive kernel, but one can nevertheless define

$$\widetilde{c}(F) := \exp\Big(-\inf\{I_K(\mu) : \mu \in \mathcal{P}(F)\}\Big).$$

The capacities c and $\widetilde{c}$ are related via the formula $1/c = \log(2/\widetilde{c})$. In particular, $\widetilde{c}(F) = 0$ if and only if $c(F) = 0$.

In this section we establish three results specific to logarithmic capacity on the circle. The first is sometimes called the maximum principle for potentials. It applies to any convex kernel, but in particular it holds for $K(t) := \log^+(2/t)$.

Theorem 2.4.2 *Let $X = \mathbb{T}$ with the chordal metric $d(z,w) := |z-w|$, and let $K : (0,\infty) \to [0,\infty)$ be a decreasing convex function. If μ is a finite positive Borel measure on $\mathbb{T}$ and $K\mu \leq M$ on* $\operatorname{supp}\mu$, *then $K\mu \leq M$ on $\mathbb{T}$.*

Proof Let I be a connected component of $\mathbb{T} \setminus \operatorname{supp}\mu$, say $I = (e^{i\alpha}, e^{i\beta})$, where $0 < \beta - \alpha \leq 2\pi$. We shall prove that $K\mu \leq M$ on I. If $\alpha \leq \gamma \leq \beta$, then

$$K\mu(e^{i\gamma}) = \int K(|e^{i\theta} - e^{i\gamma}|)\, d\mu(e^{i\theta}) = \int_{[\beta,\alpha+2\pi]} K\Big(2\sin\frac{\theta-\gamma}{2}\Big)\, d\mu(\theta).$$

Now sin is a concave function on $[0,\pi]$, and K is convex and decreasing on $[0,\infty)$. Hence, for each $\theta \in [\beta, \alpha+2\pi]$, the function $\gamma \mapsto K(2\sin(\theta-\gamma)/2))$ is convex on $[\alpha,\beta]$ (perhaps infinite at the endpoints). Integrating with respect to μ, we deduce that $\gamma \mapsto K\mu(e^{i\gamma})$ is convex on $[\alpha,\beta]$. In particular, we have $K\mu(e^{i\gamma}) \leq \max\{K\mu(e^{i\alpha}), K\mu(e^{i\beta})\}$ for all $\gamma \in (\alpha,\beta)$. Now $K\mu \leq M$ at $e^{i\alpha}, e^{i\beta}$, because both these points belong to $\operatorname{supp}\mu$. Therefore $K\mu(e^{i\gamma}) \leq M$ for all $\gamma \in (\alpha,\beta)$. In other words $K\mu \leq M$ on I, which is what we wanted to prove. □

Combining this result with Theorem 2.2.3, we obtain the following corollary.

Corollary 2.4.3 *Let $X = \mathbb{T}$ with the chordal metric $d(z,w) := |z-w|$, and let $K : (0,\infty) \to [0,\infty)$ be a decreasing convex function. Let F be a compact subset of $\mathbb{T}$ such that $c_K(F) > 0$. Let ν be an equilibrium measure for F. Then:*

(i) $K\nu(x) \le I_K(\nu)$ *for all* $x \in \mathbb{T}$,
(ii) $K\nu(x) = I_K(\nu)$ *for all* $x \in F \setminus E$, *where* $c_K^*(E) = 0$. □

The second result that we shall need is a formula for the energy of a measure μ in terms of its Fourier coefficients $\widehat{\mu}(k)$, where

$$\widehat{\mu}(k) := \int_{\mathbb{T}} e^{-ikt}\, d\mu(e^{it}) \qquad (k \in \mathbb{Z}).$$

Theorem 2.4.4 *Let $X = \mathbb{T}$ with the chordal metric $d(z,w) := |z-w|$ and let $K(t) := \log^+(2/t)$. If μ is a finite positive Borel measure on $\mathbb{T}$, then*

$$I_K(\mu) = \sum_{k\ge1} \frac{|\widehat{\mu}(k)|^2}{k} + \mu(\mathbb{T})^2 \log 2.$$

Proof We need to prove that

$$\iint \log \frac{1}{|e^{it}-e^{is}|}\, d\mu(e^{is})\, d\mu(e^{it}) = \sum_{k\ge1} \frac{|\widehat{\mu}(k)|^2}{k}. \tag{2.4}$$

Let $0 < r < 1$. Then

$$\begin{aligned}
\iint \log \frac{1}{|e^{it}-re^{is}|}\, d\mu(e^{is})\, d\mu(e^{it}) &= \operatorname{Re} \iint -\log(1-re^{i(s-t)})\, d\mu(e^{is})\, d\mu(e^{it}) \\
&= \operatorname{Re} \iint \sum_{k\ge1} \frac{r^k e^{ik(s-t)}}{k}\, d\mu(e^{is})\, d\mu(e^{it}) \\
&= \operatorname{Re} \sum_{k\ge1} \frac{r^k}{k} \overline{\widehat{\mu}(k)}\widehat{\mu}(k) \\
&= \sum_{k\ge1} \frac{r^k|\widehat{\mu}(k)|^2}{k}.
\end{aligned}$$

The result follows by letting $r \to 1^-$ on both sides. The passage of the limit inside the integral is justified as follows. If the left-hand side of (2.4) is finite, then we may use the dominated convergence theorem, exploiting the fact that $1/|e^{it}-re^{is}| \le 2/|e^{it}-e^{is}|$ for all $r < 1$ (see Lemma 1.6.7). If the left-hand side of (2.4) is infinite, then instead we use Fatou's lemma. □

The third result that we need is an estimate for logarithmic capacity in terms of Lebesgue measure on the circle. In what follows, we write $|E|$ for the arc-length measure of E.

Theorem 2.4.5 *Let E be a Borel subset of $\mathbb{T}$ with $|E| > 0$. Then*

$$c(E) \geq \frac{1}{\log(2\pi e/|E|)}.$$

Proof Let $K(t) := \log^+(2/t)$. Let F be a compact subset of $\mathbb{T}$ with $|F| > 0$. Let μ be the probability measure on F defined by $\mu(S) := |F \cap S|/|F|$. Then

$$K\mu(z) = \frac{1}{|F|}\int_{e^{i\theta}\in F} \log\frac{2}{|z - e^{i\theta}|}\,d\theta \qquad (z \in \mathbb{T}).$$

For a fixed value of $z \in \mathbb{T}$, the integral is increased by replacing F with an arc in $\mathbb{T}$ of the same length $|F|$, centered at z. Thus

$$\begin{aligned} K\mu(z) &\leq \frac{1}{|F|}\int_{-|F|/2}^{|F|/2} \log\frac{2}{|2\sin(\theta/2)|}\,d\theta \\ &\leq \frac{1}{|F|}\int_{-|F|/2}^{|F|/2} \log\frac{\pi}{|\theta|}\,d\theta \\ &= \frac{2\pi}{|F|}\int_0^{|F|/2\pi} \log(1/t)\,dt \\ &= \log(2\pi e/|F|). \end{aligned}$$

It follows that $I_K(\mu) \leq \log(2\pi e/|F|)$, and therefore $c_K(F) \geq 1/\log(2\pi e/|F|)$. This proves the result for compact sets.

The result for a general Borel set E now follows easily using the inner regularity of Lebesgue measure, namely $|E| = \sup\{|F| : F \subset E,\ F \text{ compact}\}$. □

Exercise 2.4.2 below provides an upper bound for the logarithmic capacity of an arc in terms of its measure. In conjunction with Theorem 2.4.5, it shows that $c(I) \asymp 1/(\log(1/|I|))$ as the length of the arc I tends to zero. However, no such bound is possible for general closed subsets of $\mathbb{T}$. Indeed, if E is the circular Cantor middle-third set, then $|E| = 0$, but also $c(E) > 0$. The latter follows from Theorem 2.3.5, applied with $K(t) = \log^+(2/t)$ and $e_n \sim 3^{-n-1}$ (i.e. $3^{n+1}e_n \to 1$ as $n \to \infty$).

Exercises 2.4

1. Let $X := \mathbb{T}$ and $K(t) := \log^+(2/t)$. Let μ_1, μ_2 be Borel probability measures on $\mathbb{T}$ such that $I_K(\mu_j) < \infty$. Show that

$$I_K(\mu_1) + I_K(\mu_2) - 2I_K\Big(\frac{\mu_1 + \mu_2}{2}\Big) = \frac{1}{2}\sum_{k\geq 1} \frac{|\widehat{\mu}_1(k) - \widehat{\mu}_2(k)|^2}{k}.$$

Use this to show that the equilibrium measure of a compact set of positive logarithmic capacity is unique.

2. Let $X := \mathbb{T}$ and $K(t) := \log^+(2/t)$.

   (i) Show that, if F is a closed subset of $\mathbb{T}$ and $\mu \in \mathcal{P}(F)$, then

$$\frac{1}{\sup_F K\mu} \leq c(F) \leq \frac{1}{\inf_F K\mu}.$$

   (ii) Deduce that, if I is closed subarc of $\mathbb{T}$, then

$$c(I) \leq \frac{1}{\log(2e/|I|)}.$$

   (iii) Show that the exact value of $c(\mathbb{T})$ is given by

$$c(\mathbb{T}) = \Big(\frac{1}{2\pi}\int_{-\pi}^{\pi} \log\frac{1}{|\sin(\theta/2)|}\,d\theta\Big)^{-1} = \frac{1}{\log 2}.$$

3. Let $X := \mathbb{T}$ and let $K(t) := 1/t^\alpha$, where $0 < \alpha < 1$. The corresponding capacity is called the *Riesz capacity of degree* α, and is denoted by c_α.

   (i) Show that, if E is a Borel subset of $\mathbb{T}$, then $c_\alpha(E) \geq (1-\alpha)(|E|/\pi)^\alpha$.

   (ii) Let E be the circular middle-third Cantor set. Show that $c_\alpha(E) = 0$ if and only if $\alpha \geq \log 2/\log 3$.

4. Again, let $X := \mathbb{T}$ and let $K(t) := 1/t^\alpha$, where $0 < \alpha < 1$. Show that, if μ is a finite positive Borel measure on $\mathbb{T}$, then

$$I_K(\mu) \asymp \sum_{k\geq 0} \frac{|\widehat{\mu}(k)|^2}{(1+k)^{1-\alpha}},$$

   where the implied constants depend only on α.

Notes on Chapter 2

§2.1

The material in this section is fairly standard. The abstract approach adopted here is inspired by the treatments in [29] and [70].

§2.2

The notion of equilibrium measure and the fundamental theorem 2.2.3 go back to the thesis of Frostman [48]. In many circumstances, the equilibrium measure is unique; for more on this, see for example [29, §III, Theorem 6] and [61, §III, Proposition 4].

§2.3

The results in this section are based on the treatments in [29] and [94], though the history of capacity of generalized Cantor sets goes back at least to the paper of Ohtsuka [87]. The formulation of Definition 2.3.4 and Theorem 2.3.5 was inspired by an article of Monterie [81].

§2.4

Logarithmic capacity may be approached in many different ways: via energy [5, 14, 57, 60, 70, 93, 106, 120], potentials [1, 6, 29, 61, 80], réduites [12, 38, 59], Green's functions [51, 65, 85] and transfinite diameter [92]. We have adopted the energy approach. Though these definitions may give rise to different numbers for the logarithmic capacity of a set, they are all equivalent in the sense that any one of them can be made small if any other is sufficiently small. In particular, the sets of logarithmic capacity zero are the same, whichever definition is used.

3

Boundary behavior

Since the Dirichlet space is contained in the Hardy space, it follows that each $f \in \mathcal{D}$ has non-tangential limits at almost every point of the unit circle, the exceptional set being of Lebesgue measure zero. In fact much more is true: according to a celebrated theorem of Beurling, the exceptional set is even of logarithmic capacity zero. Our primary goal in this chapter is to prove this result, which will play an important role in what follows. Our proof will be based on a representation formula for functions in the Dirichlet space, which is of interest in its own right. Using this formula, we shall deduce not only Beurling's theorem, but also two other results: the so-called strong-type inequality for capacity, and a theorem on convergence in exponentially tangential approach regions.

3.1 The Cauchy transform

We begin with some notation.

Definition 3.1.1 Let $\mathbb{A} := \{z \in \mathbb{C} : 1 < |z| < 2\}$. We write $L^2(\mathbb{A})$ for the Hilbert space of measurable functions $g : \mathbb{A} \to \mathbb{C}$ such that

$$\|g\|_{L^2(\mathbb{A})}^2 := \frac{1}{\pi} \int_{\mathbb{A}} |g(w)|^2 \, dA(w) < \infty.$$

Given $g \in L^2(\mathbb{A})$, we define its *Cauchy transform* $Cg : \mathbb{D} \to \mathbb{C}$ by

$$Cg(z) := \frac{1}{\pi} \int_{\mathbb{A}} \frac{g(w)}{w - z} \, dA(w) \qquad (z \in \mathbb{D}).$$

Using the Cauchy–Schwarz inequality, we see that $Cg(z)$ is well defined for all $z \in \mathbb{D}$, and Cg is holomorphic in $\mathbb{D}$. In fact, as the next result shows, it belongs to the Dirichlet space.

Theorem 3.1.2 *If $g \in L^2(\mathbb{A})$, then $Cg \in \mathcal{D}$ and $\|Cg\|_{\mathcal{D}} \le (3/2)^{1/2}\|g\|_{L^2(\mathbb{A})}$.*

Proof Let $g \in L^2(\mathbb{A})$. For each $z \in \mathbb{D}$, we have

$$Cg(z) = \sum_{k\ge0}\Big(\frac{1}{\pi}\int_{\mathbb{A}}\frac{g(w)}{w^{k+1}}\,dA(w)\Big)z^k = \sum_{k\ge0}\langle g,\phi_k\rangle_{L^2(\mathbb{A})}z^k, \tag{3.1}$$

where $\phi_k(w) := 1/\overline{w}^{k+1}$, and $\langle\cdot,\cdot\rangle_{L^2(\mathbb{A})}$ denotes the inner product on $L^2(\mathbb{A})$. Hence

$$\|Cg\|_{\mathcal{D}}^2 = \sum_{k\ge0}(k+1)|\langle g,\phi_k\rangle_{L^2(\mathbb{A})}|^2.$$

Now $(\phi_k)_{k\ge0}$ is an orthogonal sequence in $L^2(\mathbb{A})$ so, by Bessel's inequality,

$$\sum_{k\ge0}(k+1)|\langle g,\phi_k\rangle_{L^2(\mathbb{A})}|^2 \le B\|g\|_{L^2(\mathbb{A})}^2,$$

where $B := \sup_{k\ge0}(k+1)\|\phi_k\|_{L^2(\mathbb{A})}^2$. A calculation gives

$$\|\phi_k\|_{L^2(\mathbb{A})}^2 = \frac{1}{\pi}\int_{\mathbb{A}}\frac{1}{|w|^{2k+2}}\,dA(w) = 2\int_1^2\frac{dr}{r^{2k+1}} = \begin{cases}\log 4, & k=0,\\ (1-4^{-k})/k, & k\ge1,\end{cases}$$

and so

$$1 \le (k+1)\|\phi_k\|_{L^2(\mathbb{A})}^2 \le 3/2 \qquad (k\ge0). \tag{3.2}$$

Therefore $B = 3/2$ and the result follows. □

In fact the Cauchy transform maps $L^2(\mathbb{A})$ *onto* $\mathcal{D}$. This is the next result, which may be viewed as a representation formula for functions in the Dirichlet space.

Theorem 3.1.3 *Given $f \in \mathcal{D}$, there exists $g \in L^2(\mathbb{A})$ such that $f = Cg$ and $\|g\|_{L^2(\mathbb{A})} \le \|f\|_{\mathcal{D}}$.*

Proof Let $f \in \mathcal{D}$, say $f(z) = \sum_{k\ge0} a_k z^k$. Let $\phi_k(w) := 1/\overline{w}^{k+1}$, as in the previous proof, and consider $g := \sum_{k\ge0}(a_k/\|\phi_k\|_{L^2(\mathbb{A})}^2)\phi_k$. The sequence (ϕ_k) is an orthogonal sequence in $L^2(\mathbb{A})$, and by (3.2) we have

$$\sum_{k\ge0}|a_k|^2/\|\phi_k\|_{L^2(\mathbb{A})}^2 \le \sum_{k\ge0}(k+1)|a_k|^2 = \|f\|_{\mathcal{D}}^2 < \infty,$$

so the series defining g converges in $L^2(\mathbb{A})$, and $\|g\|_{L^2(\mathbb{A})} \le \|f\|_{\mathcal{D}}$. Furthermore, by (3.1), we have

$$C\phi_k(z) = \sum_{j\ge0}\langle\phi_k,\phi_j\rangle_{L^2(\mathbb{A})}z^j = \|\phi_k\|_{L^2(\mathbb{A})}^2 z^k \qquad (z\in\mathbb{D},\ k\ge0),$$

and therefore

$$Cg(z) = \sum_{k\geq 0}(a_k/\|\phi_k\|^2_{L^2(\mathbb{A})})C\phi_k(z) = \sum_{k\geq 0} a_k z^k = f(z) \qquad (z \in \mathbb{D}).$$

This completes the proof. □

Exercises 3.1

1. Show that, if $g(w) = 1/\overline{w}^2$, then $Cg(z) = (3/4)z$. Deduce that the constant $\sqrt{3}/2$ in Theorem 3.1.2 is sharp.
2. Show that $Cg = 0$ whenever $g(w) = 1/w^k$ $(k \geq 1)$.
3. Let $0 < \alpha < 1$ and let $L^2_\alpha(\mathbb{A})$ to be the set of measurable functions $g : \mathbb{A} \to \mathbb{C}$ such that

$$\|g\|^2_{L^2_\alpha(\mathbb{A})} := \frac{1}{\pi}\int_{\mathbb{A}} |g(w)|^2(|w|^2 - 1)^{-\alpha}\,dA(w) < \infty.$$

Show that the Cauchy transform is a bounded linear map $C : L^2_\alpha(\mathbb{A}) \to \mathcal{D}_\alpha$, and prove that it is surjective.

3.2 Beurling's theorem

Recall that $c(E)$ denotes the logarithmic capacity of $E \subset \mathbb{T}$, and that $c^*(E)$ is the corresponding outer capacity. Our main goal in this section is to prove the following theorem.

Theorem 3.2.1 (Beurling's theorem) *Let $f \in \mathcal{D}$. Then there exists $E \subset \mathbb{T}$ with $c^*(E) = 0$ such that, if $\zeta \in \mathbb{T} \setminus E$, then $f^*(\zeta) := \lim_{r\to 1^-} f(r\zeta)$ exists, and $f(z) \to f^*(\zeta)$ as $z \to \zeta$ inside each region $|z - \zeta| < \kappa(1 - |z|)$.*

A property is said to hold *quasi-everywhere* (q.e.) on $\mathbb{T}$, if it holds everywhere on $\mathbb{T} \setminus E$ where $c^*(E) = 0$. Thus Beurling's theorem can be summarized by saying that each $f \in \mathcal{D}$ has non-tangential limits quasi-everywhere on $\mathbb{T}$.

For the proof of Beurling's theorem, it is helpful to extend the Cauchy transform to the unit circle as follows.

Definition 3.2.2 Given $g \in L^2(\mathbb{A})$, we define its *maximal Cauchy transform* $\widetilde{C}g : \mathbb{T} \to [0, \infty]$ by

$$\widetilde{C}g(\zeta) := \frac{1}{\pi}\int_{\mathbb{A}} \frac{|g(w)|}{|w - \zeta|}\,dA(w) \qquad (\zeta \in \mathbb{T}).$$

If $\zeta \in \mathbb{T}$ and $\widetilde{C}g(\zeta) < \infty$, then we set

$$Cg(\zeta) := \frac{1}{\pi} \int_{\mathbb{A}} \frac{g(w)}{w - \zeta}\, dA(w).$$

Theorem 3.2.3 *Let $g \in L^2(\mathbb{A})$. Then $\widetilde{C}g$ is lower semicontinuous on $\mathbb{T}$.*

Proof Let $\zeta_n \to \zeta_0$ in $\mathbb{T}$. By Fatou's lemma,

$$\liminf_{n\to\infty} \frac{1}{\pi} \int_{\mathbb{A}} \frac{|g(w)|}{|w - \zeta_n|}\, dA(w) \geq \frac{1}{\pi} \int_{\mathbb{A}} \frac{|g(w)|}{|w - \zeta_0|}\, dA(w).$$

In other words, $\liminf_{n\to\infty} \widetilde{C}g(\zeta_n) \geq \widetilde{C}g(\zeta_0)$, as required. □

The next theorem shows that $\widetilde{C}g$ plays the role of a sort of maximal function.

Theorem 3.2.4 *Let $g \in L^2(\mathbb{A})$, let $\zeta \in \mathbb{T}$, and suppose that $\widetilde{C}g(\zeta) < \infty$. Then*

$$|Cg(z)| \leq \Bigl(1 + \frac{|z - \zeta|}{1 - |z|}\Bigr)\widetilde{C}g(\zeta) \qquad (z \in \mathbb{D}), \tag{3.3}$$

and $Cg(z) \to Cg(\zeta)$ as $z \to \zeta$ in each region $|z - \zeta| < \kappa(1 - |z|)$.

Proof For $z \in \mathbb{D}$, we have

$$|Cg(z)| \leq \frac{1}{\pi} \int_{\mathbb{A}} \frac{|g(w)|}{|w - z|}\, dA(w) \leq \Bigl(\sup_{w\in\mathbb{A}} \frac{|w - \zeta|}{|w - z|}\Bigr)\widetilde{C}g(\zeta).$$

Now, if $z \in \mathbb{D}$ and $w \in \mathbb{A}$, then

$$\frac{|w - \zeta|}{|w - z|} \leq \frac{|w - z| + |z - \zeta|}{|w - z|} \leq 1 + \frac{|z - \zeta|}{|w - z|} \leq 1 + \frac{|z - \zeta|}{1 - |z|}.$$

This gives (3.3).

Turning now to the second part of the theorem, for $\delta > 0$ let us define $g_\delta(w) := g(w)1_{\{1<|w|<1+\delta\}}$. Then, for $z \in \mathbb{D}$, we have

$$|Cg(z) - Cg(\zeta)| \leq |C(g - g_\delta)(z) - C(g - g_\delta)(\zeta)| + |Cg_\delta(z)| + |Cg_\delta(\zeta)|.$$

Now $C(g - g_\delta)$ is holomorphic on a neighborhood of $\overline{\mathbb{D}}$, so in particular

$$\lim_{\substack{z\to\zeta \\ z\in\mathbb{D}}} C(g - g_\delta)(z) = C(g - g_\delta)(\zeta).$$

Also, by (3.3),

$$|Cg_\delta(z)| \leq \Bigl(1 + \frac{|z - \zeta|}{1 - |z|}\Bigr)\widetilde{C}g_\delta(\zeta),$$

and clearly $|Cg_\delta(\zeta)| \leq \widetilde{C}g_\delta(\zeta)$. Putting these facts together, we deduce that, for each $\kappa > 0$,

$$\limsup_{\substack{z\to\zeta \\ |z-\zeta|\leq\kappa(1-|z|)}} |Cg(z) - Cg(\zeta)| \leq (\kappa + 2)\widetilde{C}g_\delta(\zeta).$$

This inequality holds for each $\delta > 0$. As $\widetilde{C}g(\zeta) < \infty$, the dominated convergence theorem implies that $\widetilde{C}g_\delta(\zeta) \to 0$ as $\delta \to 0$. Hence, for each $\kappa > 0$,

$$\limsup_{\substack{z\to\zeta \\ |z-\zeta|\le\kappa(1-|z|)}} |Cg(z) - Cg(\zeta)| = 0.$$

This gives the result. □

Theorem 3.2.4 begs the question: on how large a set can $\widetilde{C}g(\zeta)$ be infinite? This is where logarithmic capacity enters the picture.

Theorem 3.2.5 *Let $g \in L^2(\mathbb{A})$. Then*

$$c(\widetilde{C}g > t) \le A\|g\|^2_{L^2(\mathbb{A})}/t^2 \qquad (t > 0),$$

where A is an absolute constant.

The key to the proof of this theorem is the following elementary lemma.

Lemma 3.2.6 *There exists $B \ge 2$ such that, for all $\zeta_1, \zeta_2 \in \mathbb{T}$,*

$$\frac{1}{\pi}\int_{\mathbb{A}} \frac{dA(w)}{|w-\zeta_1||w-\zeta_2|} \le 2\log\frac{B}{|\zeta_1-\zeta_2|}.$$

Proof Making the change of variable $z := (w-\zeta_1)/(\zeta_2-\zeta_1)$, we obtain

$$\int_{1<|w|<2} \frac{dA(w)}{|w-\zeta_1||w-\zeta_2|} = \int_{1<|\zeta_1+(\zeta_2-\zeta_1)z|<2} \frac{dA(z)}{|z||z-1|}$$
$$< \int_{|z|<3/|\zeta_2-\zeta_1|} \frac{dA(z)}{|z||z-1|}.$$

Now

$$\int_{|z|<2} \frac{dA(z)}{|z||z-1|} \le \int_{|z|<2} \Big(\frac{1}{|z|} + \frac{1}{|z-1|}\Big)\,dA(z)$$
$$\le \int_{|z|<2} \frac{dA(z)}{|z|} + \int_{|z|<3} \frac{dA(z)}{|z|}$$
$$= 4\pi + 6\pi.$$

Also, for $R > 2$,

$$\int_{2\le|z|<R} \frac{dA(z)}{|z||z-1|} \le 2\pi\int_2^R \frac{dr}{r-1} \le 2\pi\log R.$$

If we put all these inequalities together, then we obtain

$$\frac{1}{\pi}\int_{\mathbb{A}} \frac{dA(w)}{|w-\zeta_1||w-\zeta_2|} \le 10 + 2\log\frac{3}{|\zeta_2-\zeta_1|}.$$

This gives the required inequality with $B = 3e^5$. □

Proof of Theorem 3.2.5 For the time being, we work with the capacity c_K, where K is the kernel $K(t) := \log^+(B/t)$, and B is the constant in Lemma 3.2.6. At the end of the proof we shall make the necessary adjustments to obtain the standard logarithmic capacity c.

Let $t > 0$ and let F be a compact subset of $\{\zeta \in \mathbb{T} : \widetilde{C}g(\zeta) > t\}$. Let μ be a Borel probability measure on F. Then, clearly,

$$\int_F \widetilde{C}g(\zeta)\, d\mu(\zeta) \geq t.$$

On the other hand, using successively Fubini's theorem, the Cauchy–Schwarz inequality, Fubini again and finally Lemma 3.2.6, we have

$$\begin{aligned}
\int_F \widetilde{C}g(\zeta)\, d\mu(\zeta) &= \frac{1}{\pi}\int_{\mathbb{A}}\int_F \frac{|g(w)|}{|w-\zeta|}\, d\mu(\zeta)\, dA(w) \\
&\leq \|g\|_{L^2(\mathbb{A})}\Big(\frac{1}{\pi}\int_{\mathbb{A}}\Big(\int_F \frac{1}{|w-\zeta|}\, d\mu(\zeta)\Big)^2 dA(w)\Big)^{1/2} \\
&= \|g\|_{L^2(\mathbb{A})}\Big(\int_F\int_F\Big(\frac{1}{\pi}\int_{\mathbb{A}} \frac{dA(w)}{|w-\zeta_1||w-\zeta_2|}\Big) d\mu(\zeta_1)\, d\mu(\zeta_2)\Big)^{1/2} \\
&\leq \|g\|_{L^2(\mathbb{A})}\Big(\int_F\int_F 2\log\frac{B}{|\zeta_1-\zeta_2|}\, d\mu(\zeta_1)\, d\mu(\zeta_2)\Big)^{1/2}.
\end{aligned}$$

Hence

$$\int_F \widetilde{C}g(\zeta)\, d\mu(\zeta) \leq \|g\|_{L^2(\mathbb{A})}(2I_K(\mu))^{1/2}. \tag{3.4}$$

If we put all this together, then we obtain $t \leq \|g\|_{L^2(\mathbb{A})}(2I_K(\mu))^{1/2}$, which, upon rearrangement, becomes $1/I_K(\mu) \leq 2\|g\|^2_{L^2(\mathbb{A})}/t^2$. As this holds for all $\mu \in \mathcal{P}(F)$, it follows that

$$c_K(F) \leq 2\|g\|^2_{L^2(\mathbb{A})}/t^2.$$

And as this holds for each compact F in $\{\widetilde{C}g > t\}$, we get

$$c_K(\widetilde{C}g > t) \leq 2\|g\|^2_{L^2(\mathbb{A})}/t^2.$$

It just remains to relate c_K to c. For this, note that the difference between their respective kernels is $\log B - \log 2 = \log(B/2)$, and so $1/c_K(F) - 1/c(F) = \log(B/2)$ for all compact sets F. Hence

$$c(F)/c_K(F) = 1 + \log(B/2)c(F) \leq 1 + \log(B/2)c(\mathbb{T}).$$

It follows that $c(\widetilde{C}g > t) \leq A\|g\|^2_{L^2(\mathbb{A})}/t^2$, where $A := 2(1 + \log(B/2)c(\mathbb{T}))$. □

Corollary 3.2.7 *If $g \in L^2(\mathbb{A})$, then $\widetilde{C}g < \infty$ quasi-everywhere on $\mathbb{T}$.*

Proof Since $\widetilde{C}g$ is lower semicontinuous, the set $\{\zeta \in \mathbb{T} : \widetilde{C}g(\zeta) > t\}$ is open in $\mathbb{T}$ for each $t > 0$. Thus $c^*(\widetilde{C}g = \infty) \leq c(\widetilde{C}g > t) \leq A\|g\|^2_{L^2(\mathbb{A})}/t^2$ for all $t > 0$. Letting $t \to \infty$, we get $c^*(\widetilde{C}g = \infty) = 0$. □

Finally, we are in a position to prove Beurling's theorem.

Proof of Theorem 3.2.1 Let $f \in \mathcal{D}$. From Theorem 3.1.3 we have $f = Cg$ for some $g \in L^2(\mathbb{A})$. By Theorem 3.2.4, Cg has non-tangential limits at every $\zeta \in \mathbb{T}$ for which $\widetilde{C}g(\zeta) < \infty$. And by Corollary 3.2.7, we have $\widetilde{C}g(\zeta) < \infty$ quasi-everywhere. The proof is complete. □

Exercises 3.2

1. The aim of this exercise is to establish an analogue of Beurling's theorem for the weighted Dirichlet spaces $\mathcal{D}_\alpha$. Fix α with $0 < \alpha < 1$. We write c_α for the Riesz capacity of degree α, introduced in Exercise 2.4.3.
   (i) Show that there exists a constant $B_\alpha > 0$ such that, for all $\zeta_1, \zeta_2 \in \mathbb{T}$,
   $$\frac{1}{\pi}\int_{\mathbb{A}} \frac{(|w|^2-1)^{-\alpha}}{|w-\zeta_1||w-\zeta_2|}\,dA(w) \leq \frac{B_\alpha}{|\zeta_1-\zeta_2|^\alpha}.$$
   [Hint: By symmetry, it is enough to estimate the integral on the set $\mathbb{A}' := \mathbb{A} \cap \{w : |w-\zeta_1| \leq |w-\zeta_2|\}$. Writing $w = \zeta_1 + re^{i\theta}$, show that the integral on $\mathbb{A}'$ is majorized by
   $$\frac{1}{\pi}\int_{r=0}^{3}\int_{\theta=0}^{2\pi} \frac{(2r|\cos\theta|)^{-\alpha}}{r\max\{r, |\zeta_1-\zeta_2|/2\}}\, r\,dr\,d\theta,$$
   and deduce the required estimate.]
   (ii) Let $L^2_\alpha(\mathbb{A})$ be the space introduced in Exercise 3.1.3. Show that there exists a constant $A_\alpha > 0$ such that, for all $g \in L^2_\alpha(\mathbb{A})$,
   $$c_\alpha(\widetilde{C}g > t) \leq A_\alpha\|g\|^2_{L^2_\alpha(\mathbb{A})}/t^2 \qquad (t > 0).$$
   (iii) Deduce that, if $f \in \mathcal{D}_\alpha$, then it has a non-tangential limit at each point of $\mathbb{T}$ outside a set of outer c_α-capacity zero.

3.3 Weak-type and strong-type inequalities

The proof of Beurling's theorem above yields a lot more information about the boundary values of functions in $\mathcal{D}$. In particular, the following important result is more or less an immediate consequence.

Theorem 3.3.1 (Weak-type inequality for capacity) *Let $f \in \mathcal{D}$. Then*

$$c^*(|f^*| > t) \leq A\|f\|_{\mathcal{D}}^2/t^2 \qquad (t > 0), \tag{3.5}$$

where A is an absolute constant.

Proof Using Theorem 3.1.3, we can find a function $g \in L^2(\mathbb{A})$ such that $f = Cg$ and $\|g\|_{L^2(\mathbb{A})} \leq \|f\|_{\mathcal{D}}$. By Theorem 3.2.3 the set $\{\widetilde{C}g > t\}$ is open in $\mathbb{T}$, and by Theorem 3.2.5 we have $c(\widetilde{C}g > t) \leq A\|g\|_{L^2(\mathbb{A})}^2/t^2$, where A is an absolute constant. Finally, $|f^*| = |Cg| \leq \widetilde{C}g$ q.e. on $\mathbb{T}$. Assembling the pieces, we obtain (3.5). □

Recall that $|\cdot|$ denotes Lebesgue measure on $\mathbb{T}$, and that it is related to logarithmic capacity via Theorem 2.4.5.

Corollary 3.3.2 *Let $f \in \mathcal{D}$. Then*

$$\Big|\{|f^*| > t\}\Big| \leq Ae^{-Bt^2/\|f\|_{\mathcal{D}}^2} \qquad (t > 0),$$

where $A, B > 0$ are absolute constants.

Proof Combine Theorems 3.3.1 and 2.4.5. □

We saw in Exercise 1.1.7 that $\mathcal{D} \subset H^p$ for all $p < \infty$. This amounts to saying that, if $f \in \mathcal{D}$, then $|f^*|^p \in L^1(\mathbb{T})$ for all $p < \infty$. The next result strengthens this considerably.

Corollary 3.3.3 *Let $f \in \mathcal{D}$. Then* $\exp(|f^*|^2) \in L^1(\mathbb{T})$.

Proof Assume first that $\|f\|_{\mathcal{D}}^2 < B$, where B is the constant in Corollary 3.3.2. Then

$$\begin{aligned}
\int_{\mathbb{T}} (e^{|f^*(e^{i\theta})|^2} - 1)\, d\theta &= \int_{\mathbb{T}} \int_0^{|f^*(e^{i\theta})|} 2te^{t^2}\, dt\, d\theta \\
&= \int_{t=0}^{\infty} \int_{\{|f^*|>t\}} 2te^{t^2}\, d\theta\, dt \\
&= \int_{t=0}^{\infty} 2te^{t^2} |\{|f^*| > t\}|\, dt \\
&\leq \int_{t=0}^{\infty} 2Ate^{t^2} e^{-Bt^2/\|f\|_{\mathcal{D}}^2}\, dt \\
&< \infty.
\end{aligned}$$

For the general case, write $f = p + g$, where p is a polynomial and $\|g\|_{\mathcal{D}}^2 < B/2$. Then $|f^*|^2 \leq 2|p|^2 + 2|g^*|^2$ q.e. on $\mathbb{T}$, whence

$$\int_{\mathbb{T}} e^{|f^*(e^{i\theta})|^2}\, d\theta \leq e^{2\|p\|_\infty^2} \int_{\mathbb{T}} e^{2|g^*(e^{i\theta})|^2}\, d\theta < \infty. \qquad \square$$

This result is close to being optimal. If $p > 2$, then there exists $f \in \mathcal{D}$ such that $\exp(|f^*|^p) \notin L^1(\mathbb{T})$ (see Exercise 3.3.3).

The weak-type inequality, Theorem 3.3.1, can be strengthened a little further. This extra strength turns out to be vital for certain applications, for example, the characterization of Carleson measures to be derived in §5.2.

Theorem 3.3.4 (Strong-type inequality for capacity) *Let $f \in \mathcal{D}$. Then*

$$\int_0^\infty c^*(|f^*| > t)\, dt^2 \le A\|f\|_{\mathcal{D}}^2, \tag{3.6}$$

where A is an absolute constant.

The strategy of the proof is the same as for the weak-type inequality. We first establish the following inequality for the Cauchy transform.

Theorem 3.3.5 *Let $g \in L^2(\mathbb{A})$. Then*

$$\int_0^\infty c(\widetilde{C}g > t)\, dt^2 \le A\|g\|_{L^2(\mathbb{A})}^2, \tag{3.7}$$

where A is an absolute constant.

Proof Once again, it will be convenient to work with the capacity c_K, where $K(t) := \log^+(B/t)$ and B is the constant in Lemma 3.2.6. We shall prove that, if $g \in L^2(\mathbb{A})$, then

$$\int_0^\infty c_K(Cg > t)\, dt^2 \le 8\|g\|_{L^2(\mathbb{A})}^2. \tag{3.8}$$

If so, then, since c is majorized by a multiple of c_K, the same inequality holds with c_K by c, and 8 replaced by another absolute constant.

Let $n \ge 1$ and, for $k = -n, \dots, n$, let F_k be a compact subset of $\{\widetilde{C}g > 2^k\}$. Set $\mu := \sum_{k=-n}^n 2^k c_K(F_k)\nu_k$, where ν_k is an equilibrium measure for F_k. Clearly,

$$\int_{\mathbb{T}} \widetilde{C}g(\zeta)\, d\mu(\zeta) \ge \sum_{k=-n}^{n} 2^{2k} c_K(F_k).$$

On the other hand, by (3.4) we have

$$\int_{\mathbb{T}} \widetilde{C}g(\zeta)\, d\mu(\zeta) \le \|g\|_{L^2(\mathbb{A})}(2I_K(\mu))^{1/2}.$$

Now, by Corollary 2.4.3, we have $K\nu_k \le I_K(\nu_k)$ on $\mathbb{T}$. Hence, for all j, k,

$$\int K\nu_j\, d\nu_k = \int K\nu_k\, d\nu_j \le \min\{I_K(\nu_j), I_K(\nu_k)\},$$

and therefore

$$\begin{aligned}
I_K(\mu) &= \sum_{k=-n}^{n} \sum_{j=-n}^{n} 2^j c_K(F_j) 2^k c_K(F_k) \int_{\mathbb{T}} K\nu_j \, d\nu_k \\
&\leq \sum_{k=-n}^{n} \sum_{j=-n}^{n} 2^j c_K(F_j) 2^k c_K(F_k) \min\{I_K(\nu_j), I_K(\nu_k)\} \\
&= \sum_{k=-n}^{n} \sum_{j=-n}^{n} 2^j 2^k \min\{c_K(F_j), c_K(F_k)\} \\
&\leq 2 \sum_{k=-n}^{n} \sum_{j=-n}^{k} 2^j 2^k c_K(F_k) \\
&\leq 4 \sum_{k=-n}^{n} 2^{2k} c_K(F_k).
\end{aligned}$$

If we put all this together, then we obtain

$$\sum_{k=-n}^{n} 2^{2k} c_K(F_k) \leq \|g\|_{L^2(\mathbb{A})} \Big(8 \sum_{k=-n}^{n} 2^{2k} c_K(F_k) \Big)^{1/2},$$

whence

$$\sum_{k=-n}^{n} 2^{2k} c_K(F_k) \leq 8 \|g\|^2_{L^2(\mathbb{A})}.$$

As this holds for all such compact sets F_k, we get

$$\sum_{k=-n}^{n} 2^{2k} c_K(\widetilde{C}g > 2^k) \leq 8 \|g\|^2_{L^2(\mathbb{A})}.$$

Finally, letting $n \to \infty$, we obtain

$$\sum_{k=-\infty}^{\infty} 2^{2k} c_K(\widetilde{C}g > 2^k) \leq 8 \|g\|^2_{L^2(\mathbb{A})},$$

and this implies (3.8). □

Proof of Theorem 3.3.4 By Theorem 3.1.3, there exists a function $g \in L^2(\mathbb{A})$ such that $f = Cg$ and $\|g\|_{L^2(\mathbb{A})} \leq \|f\|_{\mathcal{D}}$. As $|f^*| = |Cg| \leq \widetilde{C}g$ q.e. on $\mathbb{T}$, we have

$$\int_0^\infty c^*(|f^*| > t) \, dt^2 \leq \int_0^\infty c(\widetilde{C}g > t) \, dt^2 \leq A \|g\|^2_{L^2(\mathbb{A})} \leq A \|f\|^2_{\mathcal{D}},$$

which establishes (3.6). □

Exercises 3.3

1. Show that the strong-type inequality (3.6) implies the weak-type inequality (3.5) (with the same constant A).
2. Show that the strong-type inequality (3.6) holds with

$$A := 8 + (40 + \log(3/2))/\log 2 < 70.$$

3. Let $p > 2$, and define

$$f(z) := \Big(\log \frac{1}{1-z}\Big)^{1/p} \qquad (z \in \mathbb{D}).$$

Show that $\mathcal{D}(f) < \infty$, but that $\exp(|f^*|^p) \notin L^1(\mathbb{T})$.

4. The aim of this exercise is to establish versions of the strong- and weak-type inequalities for the weighted Dirichlet spaces $\mathcal{D}_\alpha$. Fix α with $0 < \alpha < 1$. We use the same notation as in Exercise 3.2.1. In particular, c_α denotes the Riesz capacity of degree α.
   (i) Show that, if $g \in L^2_\alpha(\mathbb{A})$, then

$$\int_0^\infty c_\alpha(\widetilde{C}g > t)\, dt^2 \le A_\alpha \|g\|^2_{L^2_\alpha(\mathbb{A})},$$

   where A_α is a positive constant depending only on α.
   (ii) Deduce that, if $f \in \mathcal{D}_\alpha$, then

$$\int_0^\infty c_\alpha^*(|f^*| > t)\, dt^2 \le \widetilde{A}_\alpha \|f\|^2_{\mathcal{D}_\alpha},$$

   where $\widetilde{A}_\alpha$ is a positive constant depending only on α.
   (iii) Deduce that, if $f \in \mathcal{D}_\alpha$, then

$$c_\alpha^*(|f^*| > t) \le \widetilde{A}_\alpha \|f\|^2_{\mathcal{D}_\alpha}/t^2 \qquad (t > 0).$$

3.4 Sharpness results

To what extent are the theorems of the preceding sections sharp? For example, in Beurling's theorem, is capacity zero the right condition for the exceptional set? Is the strong-type inequality really the best result of its kind? In this section we address these questions by proving partial converses to Beurling's theorem and the strong-type inequality.

We begin with a converse to Beurling's theorem.

Theorem 3.4.1 *Let E be a closed subset of $\mathbb{T}$ of logarithmic capacity zero. Then there exists $f \in \mathcal{D}$ such that $\lim_{z\to\zeta} \operatorname{Re} f(z) = \infty$ for all $\zeta \in E$. The function f may be chosen to be continuous on $\overline{\mathbb{D}} \setminus E$.*

For the proof, we need a lemma describing the functions that will be used as the basic building blocks in the proof of the theorem.

Lemma 3.4.2 *Let $K(t) := \log^+(2/t)$ and let μ be a probability measure on $\mathbb{T}$ with $I_K(\mu) < \infty$. Define*

$$f_\mu(z) := \int \log\Big(\frac{2}{1 - ze^{-it}}\Big)\, d\mu(e^{it}) \qquad (z \in \mathbb{D}).$$

Then:

(i) $f_\mu \in \mathcal{D}$ *and* $\mathcal{D}(f_\mu) = I_K(\mu) - \log 2$,
(ii) $|\operatorname{Im} f_\mu(z)| \le \pi/2$ *on* $\mathbb{D}$,
(iii) $0 \le \operatorname{Re} f_\mu(z) \le \log(2/\operatorname{dist}(z, \operatorname{supp}\mu))$ *on* $\mathbb{D}$,
(iv) $\operatorname{Re} f_\mu^*(\zeta) \ge K\mu(\zeta)$ *q.e. on* $\mathbb{T}$.

Proof (i) Clearly f_μ is holomorphic in $\mathbb{D}$, with Taylor expansion given by

$$f_\mu(z) = \log 2 + \sum_{k\ge1} \frac{\widehat{\mu}(k)}{k} z^k.$$

Hence $\mathcal{D}(f_\mu) = \sum_{k\ge1} |\widehat{\mu}(k)|^2/k = I_K(\mu) - \log 2$, where the last equality comes from Theorem 2.4.4. In particular $\mathcal{D}(f_\mu) < \infty$, so $f_\mu \in \mathcal{D}$.

(ii) If $|w| < 1$, then the argument of $1 - w$ lies between $-\pi/2$ and $\pi/2$. Consequently, if $z \in \mathbb{D}$ and $e^{it} \in \mathbb{T}$, then the imaginary part of $\log(2/(1 - ze^{-it}))$ lies between $-\pi/2$ and $\pi/2$. The result follows easily from this.

(iii) Taking real parts, we have

$$\operatorname{Re} f(z) = \int \log \frac{2}{|e^{it} - z|}\, d\mu(e^{it}).$$

The double inequality follows easily from this.

(iv) Let $\zeta \in \mathbb{T}$ be a point where $f^*(\zeta)$ exists. Letting $z \to \zeta$ in the previous equation, and using Fatou's lemma, we get

$$\operatorname{Re} f^*(\zeta) \ge \int \log \frac{2}{|e^{it} - \zeta|}\, d\mu(e^{it}) = K\mu(\zeta). \qquad \square$$

Proof of Theorem 3.4.1 Throughout the proof, we work with the logarithmic kernel $K(t) = \log^+(2/t)$ and the corresponding capacity $c(\cdot)$.

By hypothesis $c(E) = 0$. Therefore, using Theorem 2.1.6, we can choose a decreasing sequence of closed neighborhoods E_n of E in $\mathbb{T}$ such that

$$\sum_n c(E_n)^{1/2} < \infty.$$

For each $n \geq 1$, let ν_n be an equilibrium measure for E_n, define f_{ν_n} as in Lemma 3.4.2, and set

$$f(z) := \sum_{n\geq 1} c(E_n) f_{\nu_n}(z) \qquad (z \in \mathbb{D}).$$

From Lemma 3.4.2 (ii) and (iii), it is clear that this series converges locally uniformly in $\mathbb{D}$. We shall show that f satisfies the conclusions of the theorem.

First, let us prove that $f \in \mathcal{D}$. By Lemma 3.4.2 (i), we have $\mathcal{D}(f_{\nu_n}) \leq I_K(\nu_n) = 1/c(E_n)$, the last equality by definition of equilibrium measure. As $f_{\nu_n}(0)$ does not change with n, it follows that $\|f_{\nu_n}\|_{\mathcal{D}} \leq A/c(E_n)^{1/2}$, for some absolute constant A. Therefore

$$\sum_n \|c(E_n) f_{\nu_n}\|_{\mathcal{D}} \leq A \sum_n c(E_n)^{1/2} < \infty, \tag{3.9}$$

so the series defining f converges in $\mathcal{D}$. In particular, $f \in \mathcal{D}$, as claimed.

Next we examine boundary values. Using Lemma 3.4.2 (iv), we see that $\operatorname{Re} f_{\nu_n}^* \geq K\nu_n$ q.e. on $\mathbb{T}$. Also, by Corollary 2.4.3, we have $K\nu_n = I_K(\nu_n) = 1/c(E_n)$ q.e. on E_n. Combining these observations, we deduce that

$$\operatorname{Re} f_{\nu_n}^* \geq 1/c(E_n) \quad \text{q.e. on } E_n.$$

Since $\operatorname{Re} f_{\nu_n}$ is a positive harmonic function on $\mathbb{D}$, it is the Poisson integral of a finite positive measure on $\mathbb{T}$, the absolutely continuous part of which is $(\operatorname{Re} f_{\nu_n}^*(\zeta))|d\zeta|/2\pi$. We have just seen that $\operatorname{Re} f_{\nu_n}^* \geq 1/c(E_n)$ q.e. (and hence a.e.) on E_n, which is a neighborhood of E. It follows that the unrestricted limits of $\operatorname{Re} f_{\nu_n}$ satisfy

$$\liminf_{z\to\zeta} \operatorname{Re} f_{\nu_n}(z) \geq 1/c(E_n) \qquad (\zeta \in E). \tag{3.10}$$

Therefore, for each $N \geq 1$,

$$\liminf_{z\to\zeta} \operatorname{Re} f(z) \geq \sum_{n=1}^{N} c(E_n) \liminf_{z\to\zeta} \operatorname{Re} f_{\nu_n}(z) \geq \sum_{n=1}^{N} 1 = N \qquad (\zeta \in E).$$

As N is arbitrary, we get

$$\lim_{z\to\zeta} \operatorname{Re} f(z) = \infty \qquad (\zeta \in E).$$

Finally, we show how to modify the construction so that, in addition, f is continuous on $\overline{\mathbb{D}} \setminus E$. Let E_n, ν_n and f_{ν_n} be exactly as before. In addition, we choose a sequence r_n increasing to 1 such that

$$\operatorname{Re} f_{\nu_n}(r_n\zeta) \geq 1/2c(E_n) \qquad (\zeta \in E).$$

Such a choice is possible by (3.10). Set $f_n(z) := f_{\nu_n}(r_n z)$ and $f := \sum_n c(E_n) f_n$. The Dirichlet norm of f_n is no larger than that of f_{ν_n} so, just as before, $f \in \mathcal{D}$. Also, for every N, we have

$$\liminf_{z \to \zeta} \operatorname{Re} f(z) \geq \sum_{n=1}^{N} c(E_n) \operatorname{Re} f_n(\zeta) \geq \sum_{n=1}^{N} 1/2 = N/2 \qquad (\zeta \in E),$$

so $\lim_{z\to\zeta} \operatorname{Re} f(z) = \infty$ for all $\zeta \in E$. Finally, the terms f_n in the sum defining f are all continuous on $\overline{\mathbb{D}}$, and the sum itself converges locally uniformly on $\overline{\mathbb{D}} \setminus E$, because, by parts (ii) and (iii) of Lemma 3.4.2,

$$|f_n(z)| = |f_{\nu_n}(r_n z)| \leq \pi/2 + \log(2/\operatorname{dist}(r_n z, E_n)) \qquad (z \in \overline{\mathbb{D}}).$$

Therefore f is continuous on $\overline{\mathbb{D}} \setminus E$. □

Now we turn to the strong-type inequality. The equivalence between (i) and (iii) in the next theorem demonstrates that, in certain circumstances at least, the strong-type inequality is sharp. The equivalence with (ii) will prove useful later, in Chapter 9, when we come to study cyclicity. We write d to denote arclength distance on $\mathbb{T}$, and $E_t := \{\zeta \in \mathbb{T} : d(\zeta, E) \leq t\}$.

Theorem 3.4.3 *Let E be a closed subset of $\mathbb{T}$, and let $\eta : (0, \pi] \to (0, \infty)$ be a continuous, decreasing function such that $\eta(0) := \lim_{t\to 0^+} \eta(t) = \infty$. The following are equivalent.*

(i) *There exists $f \in \mathcal{D}$ such that*

$$\liminf_{z\to\zeta} |f(z)| \geq \eta(d(\zeta, E)) \qquad (\zeta \in \mathbb{T}).$$

(ii) *For each $\epsilon > 0$, there exists $f \in \mathcal{D}$ such that $|\operatorname{Im} f| < \epsilon$ on $\mathbb{D}$ and*

$$\liminf_{z\to\zeta} \operatorname{Re} f(z) \geq \eta(d(\zeta, E)) \qquad (\zeta \in \mathbb{T}).$$

(iii) *The function η satisfies*

$$\int_0^{\pi} c(E_t)\, |d\eta^2(t)| < \infty. \tag{3.11}$$

Proof The implication (ii)⇒(i) is obvious.

For the implication (i)⇒(iii), we observe that, if f satisfies (i), then necessarily $|f^*| \geq \eta(t)$ q.e. on E_t, so

$$\int_0^\pi c(E_t)\,|d\eta^2(t)| \leq \int_0^\pi c^*(|f^*| \geq \eta(t))\,|d\eta^2(t)| = \int_{\eta(\pi)}^\infty c^*(|f^*| \geq s)\,ds^2.$$

The last integral is finite by the strong-type inequality, Theorem 3.3.4.

Finally, we turn to the implication (iii)⇒(ii). The construction of the function f follows the same general lines as in the proof of Theorem 3.4.1, but instead of using the relatively crude estimate (3.9), we exploit the Hilbert-space structure of $\mathcal{D}$ with the aid of the following lemma.

Lemma 3.4.4 *Let $(h_n)_{n\geq 1}$ be a sequence of vectors in a Hilbert space $(\mathcal{H}, \|\cdot\|)$ such that $(h_n - h_m) \perp h_m$ whenever $n \geq m$. Then $\sum_{n\geq 1} h_n/\|h_n\|^2$ converges in $\mathcal{H}$ if and only if $\sum_{n\geq 1} n/\|h_n\|^2 < \infty$.*

Proof We have $\langle h_n, h_m \rangle = \|h_m\|^2$ whenever $n \geq m$, and hence

$$\Big\|\sum_{k=m}^n \frac{h_k}{\|h_k\|^2}\Big\|^2 = \sum_{k=m}^n \sum_{l=m}^n \frac{\|h_{\min\{k,l\}}\|^2}{\|h_k\|^2\|h_l\|^2} = \sum_{k=m}^n \frac{2k - 2m + 1}{\|h_k\|^2}.$$

Thus, if $\sum_k k/\|h_k\|^2$ converges, then the partial sums of $\sum_k h_k/\|h_k\|^2$ form a Cauchy sequence, and therefore converge in $\mathcal{H}$. Conversely, if $\sum_k h_k/\|h_k\|^2$ converges in $\mathcal{H}$, then its partial sums are bounded in norm, so the calculation above (with $m = 1$) shows that the partial sums of $\sum_k k/\|h_k\|^2$ are bounded, whence $\sum_k k/\|h_k\|^2 < \infty$. □

We now return to the proof of the implication (iii)⇒(ii) in Theorem 3.4.3. Let n_0 be a positive integer with $n_0 \geq \eta(\pi)$. For $n \geq n_0$, set $\delta_n := \eta^{-1}(n)$ and $c_n := c(E_{\delta_n})$. We then have

$$\int_0^\pi c(E_t)\,|d\eta^2(t)| \geq \sum_{n>n_0} \int_{\delta_n}^{\delta_{n-1}} c(E_{\delta_n})\,|d\eta^2(t)| = \sum_{n>n_0} c_n(n^2 - (n-1)^2),$$

so condition (iii) implies that

$$\sum_{n\geq n_0} nc_n < \infty. \tag{3.12}$$

Increasing n_0, if necessary, we can further suppose that

$$\sum_{n\geq n_0} c_n < 2\epsilon/\pi. \tag{3.13}$$

For $n \geq n_0$, let ν_n be an equilibrium measure for E_{δ_n}, and let f_{ν_n} be defined as in Lemma 3.4.2. Finally, define

$$f(z) := n_0 + \sum_{n\geq n_0} c_n f_{\nu_n}(z) \qquad (z \in \mathbb{D}).$$

Clearly this series converges locally uniformly in $\mathbb{D}$. We shall show that f satisfies all the requirements of (ii).

We begin by proving that $f \in \mathcal{D}$. It will be convenient to work with the norm $\|\cdot\|$ defined by $\|h\|^2 := |h(0)|^2 + \mathcal{D}(h)$, which is equivalent to the usual norm $\|\cdot\|_{\mathcal{D}}$ on $\mathcal{D}$. Let $h_n := f_{\nu_n} - f_{\nu_n}(0) = f_{\nu_n} - \log 2$. Using Lemma 3.4.2, we have

$$\|h_n\|^2 = \mathcal{D}(f_{\nu_n}) = \iint \log \frac{1}{|\lambda - \zeta|}\, d\nu_n(\lambda)\, d\nu_n(\zeta).$$

By polarization, the associated inner product $\langle \cdot, \cdot \rangle$ satisfies

$$\begin{aligned}\langle h_n - h_m, h_m \rangle &= \iint \log \frac{1}{|\lambda - \zeta|}\, d(\nu_n - \nu_m)(\lambda)\, d\nu_m(\zeta) \\ &= \int K\nu_m\, d\nu_n - \int K\nu_m\, d\nu_m,\end{aligned}$$

where, as usual, $K(t) := \log^+(2/t)$. By Corollary 2.4.3, we have $K\nu_m = 1/c_m$ q.e. on E_{δ_m}. Hence, if $n \geq m$, then $\int K\nu_m\, d\nu_n = 1/c_m = \int K\nu_m\, d\nu_m$, and consequently $\langle h_n - h_m, h_m \rangle = 0$. Lemma 3.4.4 thus applies. Now

$$\|h_n\|^2 = I_K(\nu_n) - \log 2 = 1/c_n - \log 2.$$

Thus (3.12) implies that $\sum_n n/\|h_n\|^2 < \infty$, and consequently, by Lemma 3.4.4, $\sum_n h_n/\|h\|^2$ converges in $\mathcal{D}$. This in turn implies that $\sum_n c_n h_n$ converges in $\mathcal{D}$, whence also $\sum_n c_n f_{\nu_n}$. Thus $f \in \mathcal{D}$, as claimed.

Next, we show that $|\operatorname{Im} f| < \epsilon$ on $\mathbb{D}$. Indeed, for all $z \in \mathbb{D}$, we have

$$|\operatorname{Im} f(z)| \leq \sum_{n \geq n_0} c_n |\operatorname{Im} f_{\nu_n}(z)| \leq \sum_{n \geq n_0} c_n \pi/2 < \epsilon,$$

where we used successively the definition of f, Lemma 3.4.2 (ii) and (3.13).

Finally we prove that $\liminf_{z \to \zeta} \operatorname{Re} f(z) \geq \eta(d(\zeta, E))$ for all $\zeta \in \mathbb{T}$. This is obvious if $d(\zeta, E)) \geq \delta_{n_0}$, since $\operatorname{Re} f \geq n_0$ everywhere in $\mathbb{D}$. So assume that $d(\zeta, E) < \delta_{n_0}$, and let N be the integer such that $\delta_{N+1} \leq d(\zeta, E) < \delta_N$. By Lemma 3.4.2 (iv), for all n we have $\operatorname{Re} f^*_{\nu_n} \geq K\nu_n = 1/c_n$ q.e. on E_{δ_n}. Hence,

$$\operatorname{Re} f^* \geq n_0 + \sum_{n=n_0}^{N} c_n \operatorname{Re} f^*_{\nu_n} \geq n_0 + \sum_{n=n_0}^{N} 1 = N + 1 \quad \text{q.e. on } E_{\delta_N}.$$

Just as in the proof of Theorem 3.4.1, as $\operatorname{Re} f$ is a positive harmonic function, it is at least as large as the Poisson integral of its radial boundary values, and we have just seen that these boundary values are at least $N + 1$ q.e. (hence a.e.) on E_{δ_N}. As ζ is an interior point of E_{δ_N}, it follows that

$$\liminf_{z \to \zeta} \operatorname{Re} f(z) \geq N + 1 = \eta(\delta_{N+1}) \geq \eta(d(\zeta, E)).$$

The proof is complete. □

Exercises 3.4

1. Use Theorem 3.4.3 to show that there exists $f \in \mathcal{D}$ such that

$$\varliminf_{z\to\zeta} \operatorname{Re} f(z) \geq \log\log\frac{1}{|1-\zeta|} \qquad (\zeta \in \mathbb{T}).$$

2. Show how to modify the construction of f in Theorem 3.4.3 so as to ensure that, in addition, f is continuous on $\overline{\mathbb{D}} \setminus E$.
3. Let E be a closed subset of $\mathbb{T}$, and let $\eta : (0,2] \to (0,\infty)$ be a decreasing convex function such that

$$\int_0^2 |E_t|\eta'(t)^2\,dt < \infty. \tag{3.14}$$

(i) Let $g(w) := -3w\eta'(\operatorname{dist}(w,E)/3)$ $(w \in \mathbb{A})$. Show that $g \in L^2(\mathbb{A})$.
(ii) Let $f := Cg + \eta(1/3)$. Show that $f \in \mathcal{D}$ and that

$$\operatorname{Re} f(z) \geq \eta(\operatorname{dist}(z,E)) \qquad (z \in \mathbb{D}). \tag{3.15}$$

[This exercise is closely related to Theorem 3.4.3. It can be shown that the hypothesis (3.14) is stronger than (3.11), but the corresponding conclusion (3.15) is valid on all of $\mathbb{D}$, not just the boundary. For more details, we refer to [46, §7].]

3.5 Exponentially tangential approach regions

In this section we shall consider limits along certain tangential approach regions. We have already seen one result of this kind. In Theorem 1.5.3, we showed that, if $f \in \mathcal{D}$, then for almost every $\zeta \in \mathbb{T}$ we have $f(z) \to f^*(\zeta)$ as $z \to \zeta$ in the oricyclic region $|z-\zeta| < \kappa(1-|z|)^{1/2}$. Using the techniques developed in this chapter, we can now prove the following much stronger version of this result, in which the approach region is exponentially tangential.

Theorem 3.5.1 *Let $f \in \mathcal{D}$. Then, for a.e. $\zeta \in \mathbb{T}$, we have $f(z) \to f^*(\zeta)$ as $z \to \zeta$ in each region*

$$|z-\zeta| < \kappa\Big(\log\frac{1}{1-|z|}\Big)^{-1} \qquad (\kappa > 0).$$

The proof is based on the representation formula Theorem 3.1.3, and on a slightly more sophisticated version of the estimate (3.3). In order to state the estimate, we need to recall the notion of maximal function.

Given $h \in L^1(\mathbb{T})$, its *Hardy–Littlewood maximal function* $Mh : \mathbb{T} \to [0, \infty]$ is defined by

$$(Mh)(\zeta) := \sup_{\delta>0} \frac{1}{2\delta} \int_{|\theta - \arg\zeta|<\delta} |h(e^{i\theta})|\, d\theta \qquad (\zeta \in \mathbb{T}).$$

Mh is lower semicontinuous, and it satisfies the weak-type inequality

$$|\{Mh > t\}| \le 6\pi \|h\|_{L^1(\mathbb{T})}/t \qquad (t > 0). \tag{3.16}$$

For more details, we refer to Appendix B.

Lemma 3.5.2 *Let $g \in L^2(\mathbb{A})$ and let $\zeta \in \mathbb{T}$. Then*

$$|Cg(z)| \le 2\widetilde{C}g(\zeta) + 2\Big((Mh_g)(\zeta)|z - \zeta| \log\Big(\frac{3}{1 - |z|}\Big)\Big)^{1/2} \qquad (z \in \mathbb{D}), \tag{3.17}$$

where

$$h_g(\zeta) := \int_1^2 |g(r\zeta)|^2 r\, dr \qquad (\zeta \in \mathbb{T}).$$

Proof Fix $z \in \mathbb{D}$. Clearly

$$|Cg(z)| \le \frac{1}{\pi} \int_{\mathbb{A}} \frac{|g(w)|}{|w - z|}\, dA(w).$$

We shall estimate the right-hand side by splitting the domain of integration into two regions. Let $\delta := |z - \zeta|$ and set $\mathbb{A}_\delta := \mathbb{A} \cap \{w : |w - \zeta| \le 2\delta\}$. For $w \in \mathbb{A} \setminus \mathbb{A}_\delta$, we have

$$\Big|\frac{w - \zeta}{w - z}\Big| \le \frac{|w - \zeta|}{|w - \zeta| - \delta} \le 2,$$

and consequently

$$\frac{1}{\pi} \int_{\mathbb{A}\setminus\mathbb{A}_\delta} \frac{|g(w)|}{|w - z|}\, dA(w) \le \frac{1}{\pi} \int_{\mathbb{A}} \frac{2|g(w)|}{|w - \zeta|}\, dA(w) = 2\widetilde{C}g(\zeta).$$

In the region $\mathbb{A}_\delta$, we apply the Cauchy–Schwarz inequality:

$$\frac{1}{\pi} \int_{\mathbb{A}_\delta} \frac{|g(w)|}{|w - z|}\, dA(w) \le \Big(\frac{1}{\pi} \int_{\mathbb{A}_\delta} |g(w)|^2\, dA(w)\Big)^{1/2} \Big(\frac{1}{\pi} \int_{\mathbb{A}_\delta} \frac{1}{|w - z|^2}\, dA(w)\Big)^{1/2}.$$

Now $\mathbb{A}_\delta$ is contained in the sector $\{1 < r < 2,\ |\theta - \arg\zeta| < \pi\delta\}$ (see Figure 3.1). Hence

$$\frac{1}{\pi} \int_{\mathbb{A}_\delta} |g(w)|^2\, dA(w) \le \frac{1}{\pi} \int_{|\theta - \arg\zeta|<\pi\delta} h_g(e^{i\theta})\, d\theta \le 2\delta (Mh_g)(\zeta).$$

Also, we have

$$\frac{1}{\pi} \int_{\mathbb{A}_\delta} \frac{1}{|w - z|^2}\, dA(w) \le \frac{1}{\pi} \int_{1-|z|<|w|<3} \frac{1}{|w|^2}\, dA(w) = 2 \log \frac{3}{1 - |z|}.$$

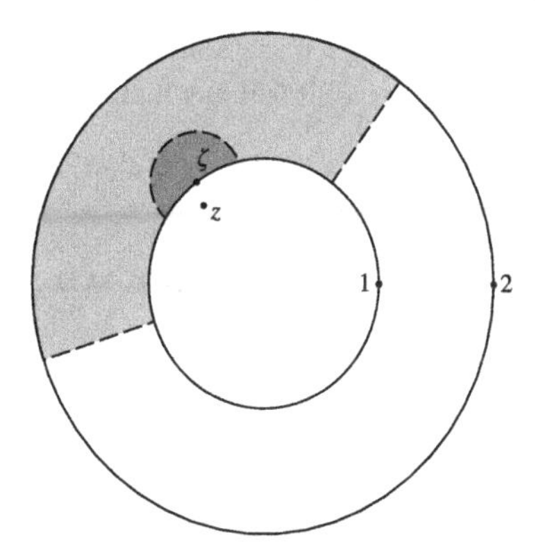

Figure 3.1 The region $\mathbb{A}_\delta$ and the encompassing sector

Putting all these estimates together, we obtain (3.17). □

Proof of Theorem 3.5.1 Let $f \in \mathcal{D}$. By Theorem 3.1.3 we can write $f = Cg$, where $g \in L^2(\mathbb{A})$. We then have $f(z) - f^*(\zeta) = Cg(z) - Cg(\zeta)$ for all $z \in \mathbb{D}$ and a.e. $\zeta \in \mathbb{T}$. For $\delta > 0$, let us set $g_\delta(w) := g(w)1_{\{1<|w|<1+\delta\}}$. Just as in the proof of Theorem 3.2.4, we have

$$|Cg(z) - Cg(\zeta)| \leq |C(g - g_\delta)(z) - C(g - g_\delta)(\zeta)| + |Cg_\delta(z)| + |Cg_\delta(\zeta)|,$$

where the first term tends to zero as $z \to \zeta$ unrestrictedly in $\mathbb{D}$, and the third term satisfies $|Cg_\delta(\zeta)| \leq \widetilde{C}g_\delta(\zeta)$. As for the second term, if we write

$$\Omega_\kappa(\zeta) := \{z \in \mathbb{D} : |z - \zeta| < \kappa/\log(3/(1 - |z|))\},$$

then by Lemma 3.5.2

$$|Cg_\delta(z)| \leq 2\widetilde{C}g_\delta(\zeta) + 2(\kappa(Mh_{g_\delta})(\zeta))^{1/2} \qquad (z \in \Omega_\kappa(\zeta)).$$

Putting this all together, we deduce that

$$\limsup_{\substack{z\to\zeta \\ z\in\Omega_\kappa(\zeta)}} |f(z) - f^*(\zeta)| \leq 3\widetilde{C}g_\delta(\zeta) + 2(\kappa(Mh_{g_\delta})(\zeta))^{1/2}.$$

Now, by Theorems 3.2.5 and 2.4.5, we have

$$|\{\widetilde{C}g_\delta > t\}| \leq Ae^{-Bt^2/\|g_\delta\|^2_{L^2(\mathbb{A})}},$$

and by (3.16)

$$|\{Mh_{g_\delta} > t\}| \leq 6\pi\|h_{g_\delta}\|_{L^1(\mathbb{T})}/t \leq A'\|g_\delta\|^2_{L^2(\mathbb{A})}/t,$$

where A, A', B are absolute constants. Also, clearly, $\|g_\delta\|_{L^2(\mathbb{A})} \to 0$ as $\delta \to 0$.

Hence, given $\epsilon > 0$, we can choose $\delta > 0$ small enough to ensure that both $|\{\widetilde{C}g_\delta > \epsilon\}| < \epsilon$ and $|\{Mh_{g_\delta} > \epsilon\}| < \epsilon$. It follows that

$$\limsup_{\substack{z\to\zeta \\ z\in\Omega_\kappa(\zeta)}} |f(z) - f^*(\zeta)| \leq 3\epsilon + 2(\kappa\epsilon)^{1/2}$$

for all $\zeta \in \mathbb{T}$ outside a set of measure 2ϵ. Letting $\epsilon \to 0$, we deduce that

$$\lim_{\substack{z\to\zeta \\ z\in\Omega_\kappa(\zeta)}} f(z) = f^*(\zeta)$$

for all $\zeta \in \mathbb{T}$ outside a set of measure zero. □

We conclude by remarking that Theorem 3.5.1 is optimal, in the sense that the tangential approach region cannot be widened any further. This statement will be made precise in Exercise 4.2.1.

Notes on Chapter 3

§3.1

The representation formula 3.1.3 is due to Dyn′kin [41]. The simple proof given here is taken from Borichev [20]. We remark that it is also possible to represent a function $f \in \mathcal{D}$ as a Cauchy-type transform of a function in $L^2(\mathbb{D})$. Indeed, if we write

$$(\overline{C}g)(z) := \frac{1}{\pi}\int_{\mathbb{D}} \frac{g(w)}{1 - \overline{w}z}\, dA(w) \qquad (z \in \mathbb{D}),$$

then $f(z) = f(0) + z\overline{C}(f')$. This formula was implicitly used in proving Theorem 1.6.6.

§3.2

Beurling's theorem (for radial limits) appeared in [16]. Our approach follows Borichev [20].

§3.3

The weak-type inequality for capacity is due to Beurling [16]. He had earlier obtained Corollary 3.3.2 in his thesis [15], showing that, if $f(0) = 0$ and $\mathcal{D}(f) \leq 1$, then for each $t > 0$ we have $|\{|f^*| > t\}| \leq 2\pi e^{1-t^2}$. This was later strengthened by Chang and Marshall [34] (see also [47] and [75]), who proved that

$$\sup\Big\{\int_{\mathbb{T}} \exp(|f^*|^2)\, d\theta : f(0) = 0,\ \mathcal{D}(f) \leq 1\Big\} < \infty.$$

The strong-type inequality for capacity is due to Hansson [54], extending earlier results of Maz′ya and of Adams. Our proof is adapted from that given in the book of Adams and Hedberg [1]. This book also contains a detailed account of the history of the strong-type inequality for capacity.

§3.4

The basic construction in the proof of Theorem 3.4.1 appears in Carleson's paper [26, p. 332]. The modification required to produce a function continuous on $\overline{\mathbb{D}} \setminus E$ is due to Brown and Cohn [23]. Theorem 3.4.3 is from [43]. Exercise 3.4.3 is based on a result in [46, §7].

§3.5

Theorem 3.5.1 is due to Nagel, Rudin and Shapiro [83]. They also showed that the result is optimal, in the sense that the tangential approach region cannot be widened any further (see Exercise 4.2.1). Our approach is an adaptation of their ideas and those in [20]. Actually this theorem is the simplest of a variety of results of this kind, with a playoff between the width of the approach region, the size of the exceptional set, and the space of functions in question. For more on this we refer to [20, 21, 82, 84, 121, 122].

4

Zero sets

If f is a function holomorphic on a domain D and $f \not\equiv 0$, then its zeros are isolated in D and of finite multiplicity. We can thus write them as a finite or infinite sequence (z_n), each zero repeated according to its multiplicity.

If f belongs to the Hardy space H^2 and $f \not\equiv 0$, then its zeros (z_n) form a Blaschke sequence; in other words, they satisfy the Blaschke condition

$$\sum_n (1 - |z_n|) < \infty. \tag{4.1}$$

Conversely, given a sequence (z_n) in $\mathbb{D}$ satisfying (4.1), there exists $f \in H^2$ whose zero set is precisely (z_n), including multiplicities. Indeed, f may even be chosen to be a Blaschke product, so that $f \in H^\infty$.

The situation for the Dirichlet space $\mathcal{D}$ is rather more complicated. Since $\mathcal{D} \subset H^2$, the Blaschke condition (4.1) is clearly necessary for (z_n) to be the zero set of some $f \in \mathcal{D}$, but it turns out not to be sufficient. Indeed, no satisfactory necessary and sufficient condition is known. The results that are known suggest that the problem is complex. In this chapter we shall present a number of the most important of these theorems. We shall also examine the distribution of zeros on the boundary.

4.1 Zero sets and uniqueness sets

Definition 4.1.1 A sequence $(z_n)_{n\geq 1}$ in $\mathbb{D}$ is a *zero set* for $\mathcal{D}$ if there exists $f \in \mathcal{D}$ such that $f(z_n) = 0$ for all n and $f(z) \neq 0$ for all other $z \in \mathbb{D}$. Repetitions are permitted. In that case, the number of times an element z_n repeats in the sequence represents the order of the zero of f at that point.

A sequence $(z_n)_{n\geq 1}$ in $\mathbb{D}$ is a *uniqueness set* for $\mathcal{D}$ if the only function $f \in \mathcal{D}$ satisfying $f(z_n) = 0$ for all n is $f \equiv 0$.

We begin with a simple class of examples of zero sets.

Theorem 4.1.2 *Let (x_n) be a sequence in $[0,1)$ such that $\sum_n(1-x_n) < \infty$. Then (x_n) is a zero set for $\mathcal{D}$.*

Proof Since the (x_n) satisfy the Blaschke condition, there exists a Blaschke product B with zeros at precisely the points (x_n). By logarithmic differentiation,

$$\frac{B'(z)}{B(z)} = \sum_n \frac{1-x_n^2}{(z-x_n)(1-x_n z)} \qquad (z \in \mathbb{D}).$$

Multiplying up by B and using Lemma 1.6.7, we have

$$|B'(z)| \le \sum_n \frac{1-x_n^2}{|1-x_n z|^2} \le \sum_n \frac{8(1-x_n)}{|1-z|^2} \qquad (z \in \mathbb{D}).$$

It follows that, if we define $f(z) := (1-z)^2 B(z)$, then f' is bounded, and in particular $f \in \mathcal{D}$. Clearly f has zeros at precisely the points (x_n). □

One might think that it would have been even simpler to take $f := B$. However, infinite Blaschke products never belong to the Dirichlet space (see Exercise 4.1.1), so multiplication by some kind of factor is necessary. The following theorem extends this idea to the case where the zeros of the Blaschke product no longer necessarily lie on a radius.

Theorem 4.1.3 *Let $f \in H^2$ and let B be a Blaschke product. Let (z_n) be the zero set of B. Then*

$$\mathcal{D}(Bf) = \mathcal{D}(f) + \sum_n \frac{1}{2\pi}\int_{\mathbb{T}} \frac{1-|z_n|^2}{|\zeta - z_n|^2}|f^*(\zeta)|^2\,|d\zeta|. \tag{4.2}$$

Proof Suppose first that $B(z) = z$. It is clear from Theorem 1.1.2 that $\mathcal{D}(zf) = \mathcal{D}(f) + \|f\|_{H^2}^2$, which gives (4.2) in this case.

Next, suppose that $B(z) = c(z_1 - z)/(1-\overline{z}_1 z)$, where $0 < |z_1| < 1$ and $|c| = 1$. Using the Möbius invariance of the Dirichlet integral, Corollary 1.4.3, we have

$$\begin{aligned}
\mathcal{D}(Bf) &= \mathcal{D}(z(f \circ B^{-1})) \\
&= \mathcal{D}(f \circ B^{-1}) + \|f \circ B^{-1}\|_{H^2}^2 \\
&= \mathcal{D}(f) + \frac{1}{2\pi}\int_{\mathbb{T}} |f^*(\zeta)|^2 |B'(\zeta)|\,|d\zeta| \\
&= \mathcal{D}(f) + \frac{1}{2\pi}\int_{\mathbb{T}} |f^*(\zeta)|^2 \frac{1-|z_1|^2}{|\zeta - z_1|^2}\,|d\zeta|.
\end{aligned}$$

This proves (4.2) in this case.

The case where B is a finite Blaschke product follows by induction.

Finally, for the general case, let us write b_n for the product of the first n terms in B, and B_n for the product of the remaining terms. By what we have already proved for finite Blaschke products, for each n we have

$$\mathcal{D}(b_n f) = \mathcal{D}(f) + \sum_{k=1}^{n} \frac{1}{2\pi} \int_{\mathbb{T}} \frac{1 - |z_k|^2}{|\zeta - z_k|^2} |f^*(\zeta)|^2 \, |d\zeta|.$$

Since $b_n f \to Bf$ locally uniformly on $\mathbb{D}$, we get $\mathcal{D}(Bf) \leq \liminf_{n\to\infty} \mathcal{D}(b_n f)$. Hence

$$\mathcal{D}(Bf) \leq \mathcal{D}(f) + \sum_{k=1}^{\infty} \frac{1}{2\pi} \int_{\mathbb{T}} \frac{1 - |z_k|^2}{|\zeta - z_k|^2} |f^*(\zeta)|^2 \, |d\zeta|.$$

For the reverse inequality, note that

$$\mathcal{D}(Bf) = \mathcal{D}(b_n B_n f) = \mathcal{D}(B_n f) + \sum_{k=1}^{n} \frac{1}{2\pi} \int_{\mathbb{T}} \frac{1 - |z_k|^2}{|\zeta - z_k|^2} |f^*(\zeta)|^2 \, |d\zeta|.$$

Here we have used the fact that $|B_n^*| = 1$ a.e. on $\mathbb{T}$. Also, since $B_n f \to f$ locally uniformly on $\mathbb{D}$, we have $\liminf_{n\to\infty} \mathcal{D}(B_n f) \geq \mathcal{D}(f)$. Hence

$$\mathcal{D}(Bf) \geq \mathcal{D}(f) + \sum_{k=1}^{\infty} \frac{1}{2\pi} \int_{\mathbb{T}} \frac{1 - |z_k|^2}{|\zeta - z_k|^2} |f^*(\zeta)|^2 \, |d\zeta|.$$

This completes the proof. □

We shall see later that this is actually a special case of a much more general result (Theorem 7.6.9). As a first application, we prove that, in the Dirichlet space, the notions of zero set and uniqueness set are complementary.

Theorem 4.1.4 *A sequence (z_n) in $\mathbb{D}$ is a zero set for $\mathcal{D}$ if and only if it is not a uniqueness set.*

Proof The 'only if' is obvious. For the 'if', suppose that (z_n) is not a uniqueness set. Then there exists $g \in \mathcal{D}$ such that $g(z_n) = 0$ for all n and $g \not\equiv 0$. The problem is that, in addition to the (z_n), there may be some additional zeros (w_n), either a finite or infinite sequence. In the case of an infinite sequence, they satisfy the Blaschke condition, because they are zeros of a function in H^2. Let B be the Blaschke product with zeros at the (w_n), and let $f := g/B$. Then $f \in H^2$ and has zeros exactly at the (z_n). Finally, by Theorem 4.1.3, we have $\mathcal{D}(f) \leq \mathcal{D}(Bf) = \mathcal{D}(g) < \infty$, and so $f \in \mathcal{D}$. □

The formula (4.2) has another interesting consequence. Recall that we write $\log^+(x) := \max\{\log(x), 0\}$.

Theorem 4.1.5 *If (z_n) is a zero set for $\mathcal{D}$, then*

$$\int_{\mathbb{T}} \log^+\Big(\sum_n \frac{1-|z_n|^2}{|\zeta - z_n|^2}\Big)\,|d\zeta| < \infty. \tag{4.3}$$

Proof By assumption, there is a function in $\mathcal{D}$ with zero set (z_n). We can factorize this function as Bf, where B is the Blaschke product whose zero set is (z_n), and $f \in H^2 \setminus \{0\}$. By Theorem 4.1.3, we have

$$\mathcal{D}(Bf) = \mathcal{D}(f) + \frac{1}{2\pi}\int_{\mathbb{T}} \sum_n \frac{1-|z_n|^2}{|\zeta - z_n|^2}|f^*(\zeta)|^2\,|d\zeta|.$$

In particular, the integral on the right-hand side is finite. Since $\log^+ x \le x$ for all $x \ge 0$, it follows that

$$\int_{\mathbb{T}} \log^+\Big(\sum_n \frac{1-|z_n|^2}{|\zeta - z_n|^2}|f^*(\zeta)|^2\Big)\,|d\zeta| < \infty.$$

Also, because $f \in H^2 \setminus \{0\}$, we necessarily have

$$\int_{\mathbb{T}} \log^+ \frac{1}{|f^*(\zeta)|^2}\,|d\zeta| < \infty.$$

As $\log^+(ab) \le \log^+ a + \log^+ b$ for all $a, b \ge 0$, we deduce that (4.3) holds. □

Although the condition (4.3) is difficult to interpret in general, it does allow us to exhibit sequences satisfying the Blaschke condition (4.1) that are uniqueness sets for $\mathcal{D}$. These sequences may even be chosen to be convergent.

Theorem 4.1.6 *Let (δ_n) be a sequence in $(0,1)$ such that $\sum_n \delta_n < \infty$ and $\sum_n \delta_n \log(1/\delta_n) = \infty$. For $n \ge 1$, set $r_n := 1 - \delta_n$ and $\theta_n := \sum_{k \ge n} \delta_k$. Then $(r_n e^{i\theta_n})$ is a uniqueness set for $\mathcal{D}$.*

Proof If $\theta \in (\theta_{n+1}, \theta_n)$, then

$$|e^{i\theta} - r_n e^{i\theta_n}| \le |\theta - \theta_n| + (1 - r_n) \le 2\delta_n,$$

and so

$$\frac{1 - r_n^2}{|e^{i\theta} - r_n e^{i\theta_n}|^2} \ge \frac{\delta_n}{4\delta_n^2} = \frac{1}{4\delta_n}.$$

Consequently, for each n, we have

$$\int_{\theta_{n+1}}^{\theta_n} \log^+\Big(\sum_k \frac{1-r_k^2}{|e^{i\theta} - r_k e^{i\theta_k}|^2}\Big)\,d\theta \ge \int_{\theta_{n+1}}^{\theta_n} \log^+\Big(\frac{1-r_n^2}{|e^{i\theta} - r_n e^{i\theta_n}|^2}\Big)\,d\theta$$

$$\ge \delta_n \log^+\Big(\frac{1}{4\delta_n}\Big).$$

Hence, finally,

$$\int_0^{2\pi} \log^+\Big(\sum_k \frac{1-r_k^2}{|e^{i\theta}-r_k e^{i\theta_k}|^2}\Big)\,d\theta \ge \sum_n \delta_n \log^+\Big(\frac{1}{4\delta_n}\Big) = \infty.$$

By Theorem 4.1.5, it follows that $(r_n e^{i\theta_n})$ is a uniqueness set. □

Comparing Theorems 4.1.2 and 4.1.6, we see that there exists a sequence (z_n) converging to 1 such that $(|z_n|)$ is a zero set but (z_n) itself is not. Thus the arguments of (z_n) play a role in determining whether it is a zero set for $\mathcal{D}$. We shall return to this topic in §4.5.

Exercises 4.1

1. Use formula (4.2) to show that no infinite Blaschke product belongs to $\mathcal{D}$.
2. Show that, in Theorem 4.1.6, we can equally well take $\theta_n := \sum_{k\ge n} \eta_k \delta_k$, where (η_k) is any positive sequence such that $\eta_k \asymp 1$ as $k \to \infty$.
3. For $n \ge 3$, let

$$z_n := \Big(1 - \frac{1}{n(\log n)^2}\Big)e^{i/\log n}.$$

Show that $(z_n)_{n\ge 3}$ is a uniqueness set for $\mathcal{D}$. [Use the previous exercise.]

4.2 Moduli of zero sets

Is there a condition on the moduli $|z_n|$ guaranteeing that the sequence (z_n) is a zero set for $\mathcal{D}$? As we have just seen, though the Blaschke condition (4.1) is necessary, it is not sufficient. A condition that *is* sufficient is given by the following theorem.

Theorem 4.2.1 *Let (z_n) be a sequence in $\mathbb{D} \setminus \{0\}$ such that*

$$\sum_n \Big\{\log\Big(\frac{1}{1-|z_n|}\Big)\Big\}^{-1} < \infty. \tag{4.4}$$

Then (z_n) is a zero set for $\mathcal{D}$.

For example, if $|z_n| = 1 - e^{-n^2}$, then (z_n) satisfies (4.4). On the other hand, if $|z_n| = 1 - e^{-n}$, then (z_n) does not satisfy (4.4), even though it does satisfy the Blaschke condition. Thus, in order to satisfy (4.4), the sequence (z_n) must approach the boundary 'very fast'.

The condition (4.4), though stringent, is optimal in the following sense.

Theorem 4.2.2 *Let (r_n) be a sequence in $(0,1)$ such that*

$$\sum_n \Big\{\log\Big(\frac{1}{1-r_n}\Big)\Big\}^{-1} = \infty. \tag{4.5}$$

Then there exists a sequence (θ_n) such that $(r_n e^{i\theta_n})$ is a uniqueness set for $\mathcal{D}$.

We now turn to the proofs of these two theorems, beginning with the second.

Proof of Theorem 4.2.2 Let $(I_n)_{n\geq 1}$ be contiguous arcs in $\mathbb{T}$ such that $|I_n| = 1/\log(1/(1-r_n))$ for all n. Let $e^{i\theta_n}$ be the midpoint of I_n for each n, and set $z_n := r_n e^{i\theta_n}$. We shall show that (z_n) is a uniqueness set for $\mathcal{D}$.

If $\zeta \in I_n$, then

$$|z_n - \zeta| \leq |z_n - e^{i\theta_n}| + |e^{i\theta_n} - \zeta| < 2\Big\{\log\frac{1}{1-r_n}\Big\}^{-1}.$$

In other words, if we set

$$\Omega(\zeta) := \Big\{z \in \mathbb{D} : |z-\zeta| < 2\Big\{\log\frac{1}{1-|z|}\Big\}^{-1}\Big\},$$

then $z_n \in \Omega(\zeta)$ for all $\zeta \in I_n$. Now, condition (4.5) implies that $\sum_n |I_n| = \infty$, so almost every $\zeta \in \mathbb{T}$ lies in infinitely many I_n. Thus, for almost every $\zeta \in \mathbb{T}$, the set $\Omega(\zeta)$ contains infinitely many (z_n). Also, by Theorem 3.5.1, if $f \in \mathcal{D}$, then for almost every $\zeta \in \mathbb{T}$, we have $f(z) \to f^*(\zeta)$ as $z \to \zeta$ in $\Omega(\zeta)$. Putting these two facts together, we see that, if $f \in \mathcal{D}$ and $f(z_n) = 0$ for all n, then $f^* = 0$ a.e. on $\mathbb{T}$, and consequently $f \equiv 0$. Thus (z_n) is indeed a uniqueness set. □

The rest of this section is devoted to the proof of Theorem 4.2.1. The strategy will be to express the problem in terms of reproducing kernels, thereby reducing it to a geometric question about Hilbert spaces. Indeed, suppose that $f \in \mathcal{D}$ and $f(z_n) = 0$ for all n but $f \not\equiv 0$. Replacing f by f/z^m, we may assume that $f(0) \neq 0$. Thus, in terms of reproducing kernels, we have $\langle f, k_{z_n}\rangle_{\mathcal{D}} = 0$ for all n and $\langle f, 1\rangle_{\mathcal{D}} \neq 0$. Obviously, for this to be possible, it is necessary that 1 lie outside the closed linear span M of $\{k_{z_n} : n \geq 1\}$. Conversely, if 1 does indeed lie outside M, then writing $f := 1 - P1$, where $P : \mathcal{D} \to M$ denotes the orthogonal projection, we have $f \not\equiv 0$ and $f \in M^\perp$, so $f(z_n) = \langle f, k_{z_n}\rangle_{\mathcal{D}} = 0$ for all n.

For the time being, we leave aside the Dirichlet space, and concentrate on the problem of estimating the distance between an element of a Hilbert space and a subspace. We shall make extensive use of the notion of positive definite and semi-definite matrices. For the definition and relevant background on this subject, see Appendix C.

Lemma 4.2.3 *Let $\mathcal{H}$ be a complex Hilbert space, and let $x_1, \ldots, x_n \in \mathcal{H}$. Let*

$$G := \begin{pmatrix} \langle x_1, x_1 \rangle & \langle x_1, x_2 \rangle & \ldots & \langle x_1, x_n \rangle \\ \vdots & \vdots & \ddots & \vdots \\ \langle x_n, x_1 \rangle & \langle x_n, x_2 \rangle & \ldots & \langle x_n, x_n \rangle \end{pmatrix}.$$

Then G is positive semi-definite. Further, it is positive definite if and only if the vectors $x_1, \ldots, x_n$ are linearly independent.

Proof Let $\lambda_1, \ldots, \lambda_n \in \mathbb{C}$. Then

$$\sum_{i,j} \langle x_i, x_j \rangle \lambda_i \overline{\lambda}_j = \langle \sum_i \lambda_i x_i, \sum_j \lambda_j x_j \rangle = \left\| \sum_i \lambda_i x_i \right\|^2 \geq 0.$$

Thus G is positive semi-definite. It fails to be positive definite if and only if there is a choice of (λ_i) not all zero such that $\| \sum_i \lambda_i x_i \| = 0$, in other words, if and only if $x_1, \ldots, x_n$ are linearly dependent vectors. □

The matrix G is called the *Gram matrix* corresponding to $x_1, \ldots, x_n$. We shall denote its determinant by $G(x_1, \ldots, x_n)$. The next result gives a distance formula in terms of Gram determinants.

Lemma 4.2.4 *Let $\mathcal{H}$ be a Hilbert space, let $x_1, \ldots, x_n \in \mathcal{H}$ be linearly independent vectors, and let $M := \operatorname{span}\{x_1, \ldots, x_n\}$. Then, for all $x \in \mathcal{H}$,*

$$\operatorname{dist}(x, M)^2 = \frac{G(x, x_1, \ldots, x_n)}{G(x_1, \ldots, x_n)}.$$

Proof Fix $x \in \mathcal{H}$, and consider the vector $y \in \mathcal{H}$ given by the formula

$$y := \begin{vmatrix} x & x_1 & \ldots & x_n \\ \langle x_1, x \rangle & \langle x_1, x_1 \rangle & \ldots & \langle x_1, x_n \rangle \\ \vdots & \vdots & \ddots & \vdots \\ \langle x_n, x \rangle & \langle x_n, x_1 \rangle & \ldots & \langle x_n, x_n \rangle \end{vmatrix}.$$

Clearly we have $\langle x, y \rangle = G(x, x_1, \ldots, x_n)$. Also, each inner product $\langle x_i, y \rangle$ is a determinant with two identical rows, and so $\langle x_i, y \rangle = 0$ $(i = 1, \ldots, n)$. Thus $y \in M^\perp$. Expanding the determinant defining y by the first row, we see that $y = G(x_1, \ldots, x_n)x + w$, where w is a linear combination of $x_1, \ldots, x_n$. Therefore, if we set $z := y/G(x_1, \ldots, x_n)$ and $t := -w/G(x_1, \ldots, x_n)$, then $x = t + z$ is a decomposition of x as a sum of a vector in M and a vector in $M^\perp$. Consequently,

$$\operatorname{dist}(x, M)^2 = \|z\|^2 = \langle x, z \rangle = \frac{\langle x, y \rangle}{G(x_1, \ldots, x_n)} = \frac{G(x, x_1, \ldots, x_n)}{G(x_1, \ldots, x_n)}.$$ □

Though this lemma gives an exact formula for the distance, the Gram determinants may be rather difficult to compute in practice, and so the following simpler estimate is useful.

Lemma 4.2.5 *Let $\mathcal{H}$ be a Hilbert space, let $x_1, \dots, x_n \in \mathcal{H}$ be linearly independent vectors such that $\langle x_i, x_j \rangle \neq 0$ for all i, j, and let $M := \operatorname{span}\{x_1, \dots, x_n\}$. Let $x \in \mathcal{H} \setminus \{0\}$, and suppose that the $n \times n$ matrix $H = (h_{ij})$, defined by*

$$h_{ij} := 1 - \frac{\langle x_i, x \rangle \langle x, x_j \rangle}{\langle x, x \rangle \langle x_i, x_j \rangle} \qquad (i, j \in \{1, \dots, n\}),$$

is positive semi-definite. Then

$$\operatorname{dist}(x, M)^2 \geq \|x\|^2 \prod_{j=1}^{n} h_{jj}.$$

Proof Let us calculate the determinant $G(x, x_1, \dots, x_n)$ using row operations. Label the rows from 0 to n, and for each $i \in \{1, \dots, n\}$, subtract $\langle x_i, x \rangle / \langle x, x \rangle$ times row 0 from row i. The resulting matrix, which of course has the same determinant, has $\langle x, x \rangle$ as its first entry, and all the other entries in the leftmost column are zero. Hence $G(x, x_1, \dots, x_n) = \langle x, x \rangle \det(A)$, where A is the $n \times n$ matrix given by

$$A_{ij} = \langle x_i, x_j \rangle - \frac{\langle x_i, x \rangle}{\langle x, x \rangle} \langle x, x_j \rangle = \langle x_i, x_j \rangle \Big(1 - \frac{\langle x_i, x \rangle \langle x, x_j \rangle}{\langle x, x \rangle \langle x_i, x_j \rangle} \Big).$$

Thus A is equal to the Hadamard product $G \circ H$, where G is the matrix defined in Lemma 4.2.3. By that lemma, G is positive definite. Also H is positive semi-definite, by assumption. It follows by Oppenheim's inequality (see Theorem C.2.3 in Appendix C) that $\det(A) \geq \det(G) \prod_{j=1}^{n} h_{jj}$. Putting this all together, we obtain

$$G(x, x_1, \dots, x_n) = \langle x, x \rangle \det(A) \geq \|x\|^2 G(x_1, \dots, x_n) \prod_{j=1}^{n} h_{jj}.$$

Combining this with the result of the Lemma 4.2.4, we obtain the desired conclusion. □

Now we return to the Dirichlet space. As indicated at the beginning of the proof, the idea is to apply the preceding theory, in particular Lemma 4.2.5, to $\mathcal{H} = \mathcal{D}$ with $x = k_0 = 1$ and $x_j = k_{z_j}$ $(j = 1, \dots, n)$, where k_{z_j} is the reproducing kernel corresponding to z_j. To be able to apply Lemma 4.2.5, we need to know that the hypotheses are satisfied. That is the subject of the next lemma.

Lemma 4.2.6 *Let $z_1, \dots, z_n$ be distinct points in $\mathbb{D} \setminus \{0\}$. Then*

(i) $k_{z_1}, \dots, k_{z_n}$ *are linearly independent,*
(ii) $\langle k_{z_i}, k_{z_j}\rangle_{\mathcal{D}} \neq 0$ *for all* i, j,
(iii) *the matrix* $H = (h_{ij})$ *defined by*

$$h_{ij} := 1 - \frac{\langle k_{z_i}, 1\rangle\langle 1, k_{z_j}\rangle}{\langle 1, 1\rangle\langle k_{z_i}, k_{z_j}\rangle} \qquad (i, j \in \{1, \dots, n\}),$$

is positive semi-definite.

For the proof, we require one more lemma.

Lemma 4.2.7 *Let* $f \in \mathrm{Hol}(\mathbb{D})$, *say* $f(z) = \sum_{n\geq 0} a_n z^n$. *Suppose that* $a_0 = 1$ *and that* $a_n > 0$ *and* $a_n^2 \leq a_{n-1}a_{n+1}$ *for all* $n \geq 1$. *Then*

$$1 - \frac{1}{f(z)} = \sum_{n\geq 0} b_n z^n \qquad (z \in \mathbb{D}), \tag{4.6}$$

where $b_n \geq 0$ *for all* n.

Proof Since $f(0) \neq 0$, the function $1 - 1/f(z)$ has a Taylor expansion $\sum_n b_n z^n$ in a neighborhood of 0. Multiplying up, we get $(\sum_n a_n z^n)(1 - \sum_n b_n z^n) = 1$ in a neighborhood of 0, and equating coefficients then gives

$$\sum_{j=0}^{n} a_{n-j} b_j = a_n \qquad (n \geq 1). \tag{4.7}$$

It follows that

$$\frac{1}{a_n}\sum_{j=0}^{n} a_{n-j} b_j = \frac{1}{a_{n-1}}\sum_{j=0}^{n-1} a_{n-j-1} b_j \qquad (n \geq 2).$$

Rearranging this last equation, and recalling that $a_0 = 1$, we get

$$b_n = \sum_{j=0}^{n-1} a_{n-j-1} b_j \Big(\frac{a_n}{a_{n-1}} - \frac{a_{n-j}}{a_{n-j-1}}\Big) \qquad (n \geq 2).$$

By assumption a_j/a_{j-1} increases with j, so the term in parentheses is non-negative. Also, we have $b_0 = 0$ and $b_1 = a_1 > 0$. By induction it follows that $b_n \geq 0$ for all n. Formula (4.7) now shows that $b_n \leq a_n$ for all n, so $\sum_n b_n z^n$ is holomorphic on $\mathbb{D}$, and the relation (4.6) holds on the whole of $\mathbb{D}$. □

Proof of Lemma 4.2.6 (i) Suppose that $\sum_{j=1}^{n} \lambda_j k_{z_j} = 0$. If $f \in \mathcal{D}$, then

$$0 = \Big\langle f, \sum_j \lambda_j k_{z_j}\Big\rangle_{\mathcal{D}} = \sum_{j=1}^{n} \overline{\lambda}_j f(z_j).$$

In particular, taking f to be a polynomial with $f(z_1) = 1$ and $f(z_j) = 0$ ($j \geq 2$), we deduce that $\lambda_1 = 0$. Likewise $\lambda_2, \dots, \lambda_n = 0$. Hence $k_{z_1}, \dots, k_{z_n}$ are linearly independent.

(ii) For all $i, j \in \{1, \dots, n\}$, we have

$$\langle k_{z_i}, k_{z_j} \rangle_{\mathcal{D}} = k_{z_i}(z_j) = \frac{1}{\overline{z}_i z_j} \log\Big(\frac{1}{1 - \overline{z}_i z_j}\Big) \neq 0.$$

(iii) For $i, j \in \{1, \dots, n\}$, we have

$$h_{ij} = 1 - \frac{\langle k_{z_i}, 1 \rangle_{\mathcal{D}} \langle 1, k_{z_j} \rangle_{\mathcal{D}}}{\langle 1, 1 \rangle_{\mathcal{D}} \langle k_{z_i}, k_{z_j} \rangle_{\mathcal{D}}} = 1 - \frac{\overline{z}_i z_j}{\log(\frac{1}{1 - \overline{z}_i z_j})}.$$

By Lemma 4.2.7, applied to $f(z) := (1/z) \log(1/(1-z)) = \sum_{n \geq 0} z^n/(n+1)$, we have

$$1 - \frac{z}{\log(\frac{1}{1-z})} = \sum_{n \geq 0} b_n z^n \qquad (z \in \mathbb{D}),$$

where $b_n \geq 0$ for all n. Therefore, given $\lambda_1, \dots, \lambda_n \in \mathbb{C}$, we have

$$\sum_{i=1}^{n} \sum_{j=1}^{n} h_{ij} \lambda_i \overline{\lambda}_j = \sum_{i=1}^{n} \sum_{j=1}^{n} \sum_{n \geq 0} b_n \overline{z}_i^n z_j^n \lambda_i \overline{\lambda}_j = \sum_{n \geq 0} b_n \Big| \sum_{i=1}^{n} \lambda_i \overline{z}_i^n \Big|^2 \geq 0.$$

This shows that H is positive semi-definite. □

Finally, we are ready to assemble all the pieces.

Proof of Theorem 4.2.1 Assume first that all the (z_n) are distinct. By Lemmas 4.2.5 and 4.2.6, for each $n \geq 1$ we have

$$\operatorname{dist}\big(1, \operatorname{span}\{k_{z_1}, \dots, k_{z_n}\}\big)^2 \geq \prod_{j=1}^{n} \Big(1 - \frac{|z_j|^2}{\log(\frac{1}{1-|z_j|^2})}\Big).$$

Letting $n \to \infty$, we deduce that

$$\operatorname{dist}(1, M)^2 \geq \prod_{j=1}^{\infty} \Big(1 - \frac{|z_j|^2}{\log(\frac{1}{1-|z_j|^2})}\Big),$$

where M denotes the closed linear span of $\{k_{z_n} : n \geq 1\}$. The condition (4.4) ensures that the infinite product converges to a positive limit, so $\operatorname{dist}(1, M) > 0$. Thus, if we set $f := 1 - P1$, where P denotes the orthogonal projection of $\mathcal{D}$ onto M, then $f \not\equiv 0$, and $f \in M^\perp$, so $f(z_n) = \langle f, k_{z_n} \rangle_{\mathcal{D}} = 0$ for all n. This proves the theorem in this case. We note, for future reference, that

$$f(0) = \langle f, 1 \rangle_{\mathcal{D}} = \langle 1 - P1, 1 \rangle_{\mathcal{D}} = \|1 - P1\|_{\mathcal{D}}^2 = \|f\|_{\mathcal{D}}^2,$$

and that

$$\|f\|_{\mathcal{D}}^2 = \operatorname{dist}(1, M)^2 \geq \prod_{j=1}^{n} \Big(1 - \frac{|z_j|^2}{\log(\frac{1}{1-|z_j|^2})}\Big).$$

There remains the case where some of the (z_n) are repeated. For this we use a small trick. Let $\epsilon > 0$, and let (w_n) be another sequence in $\mathbb{D}$, chosen so that

- $|w_n| = |z_n|$ for all n,
- $|w_n - z_n| < \epsilon$ for all n,
- the (w_n) are distinct.

By the argument in the preceding paragraph, there exists $f_\epsilon \in \mathcal{D} \setminus \{0\}$ such that $f_\epsilon(w_n) = 0$ for all n. The argument shows, moreover, that f_ϵ can be chosen so that

$$f_\epsilon(0) = \|f_\epsilon\|_{\mathcal{D}}^2 \geq \prod_{j=1}^{\infty}\Bigl(1 - \frac{|w_j|^2}{\log(\frac{1}{1-|w_j|^2})}\Bigr) = \prod_{j=1}^{n}\Bigl(1 - \frac{|z_j|^2}{\log(\frac{1}{1-|z_j|^2})}\Bigr).$$

Thus, if we set $g_\epsilon := f_\epsilon / f_\epsilon(0)$, then $\sup_{\epsilon>0} \|g_\epsilon\|_{\mathcal{D}} < \infty$, and of course $g_\epsilon(w_n) = 0$ for all n and $g_\epsilon(0) = 1$. Using Theorem 1.2.1, we see that the family $(g_\epsilon)_{\epsilon>0}$ is locally uniformly bounded, and therefore a normal family. Hence there is a sequence of ϵ tending to zero such that the corresponding functions g_ϵ converge locally uniformly on $\mathbb{D}$. The limit function g is holomorphic, and by Hurwitz's theorem $g(z_n) = 0$ for all n, even including multiplicities. Further $g(0) = 1$, so $g \not\equiv 0$. Finally, by Fatou's lemma, $\mathcal{D}(g) \leq \sup_\epsilon \mathcal{D}(g_\epsilon) < \infty$, so $g \in \mathcal{D}$. This completes the proof. □

Another proof of this theorem will be outlined in Exercise 5.4.3.

Exercises 4.2

1. [This exercise shows that the approach region in Theorem 3.5.1 is optimal.] Let $\psi : (0, 1) \to (0, \infty)$ be a function such that $\psi(t) \log(1/t) \to \infty$ as $t \to 0^+$.
   (i) Construct a sequence (r_n) in $(0, 1)$ such that
   $$\sum_n \Bigl\{\log \frac{1}{1 - r_n}\Bigr\}^{-1} < \infty \quad \text{but} \quad \sum_n \psi(1 - r_n) = \infty.$$
   (ii) Let $(I_n)_{n\geq 1}$ be contiguous arcs in $\mathbb{T}$ with $|I_n| = \psi(1 - r_n)$ for all n, let $e^{i\theta_n}$ be the midpoint of I_n and set $z_n := r_n e^{i\theta_n}$. Show that, for almost all $\zeta \in \mathbb{T}$, the set
   $$\Omega(\zeta) := \{z : |z - \zeta| < \psi(1 - |z|)\}$$
   contains infinitely many (z_n).
   (iii) Show that there exists $f \in \mathcal{D}$ with zero set (z_n).
   (iv) Deduce that the limit $\lim_{\substack{z\to\zeta \\ z\in\Omega(\zeta)}} f(z)$ fails to exist for almost all $\zeta \in \mathbb{T}$.
2. Prove that the coefficients (b_n) in Lemma 4.2.7 satisfy $\sum_{n\geq 0} b_n \leq 1$.

3. Let (z_n) be a Blaschke sequence in $\mathbb{D}$. Define

$$\mathcal{E} := \Big\{\zeta \in \mathbb{T} : \sum_n \frac{1 - |z_n|^2}{|\zeta - z_n|^2} < \infty\Big\}.$$

   (i) Show that, if (z_n) is a zero set for $\mathcal{D}$, then $\mathcal{E}$ is of measure zero.
   (ii) Show that, if (z_n) satisfies the condition (4.4), then $\mathcal{E}$ is of logarithmic capacity zero.

   [Remark: There exist sequences (z_n) that are zero sets for $\mathcal{D}$, but for which $\mathcal{E}$ is of positive capacity. See the notes at the end of the chapter.]

4.3 Boundary zeros I: sets of capacity zero

We know that, if $f \in \mathcal{D}$, then f^* exists everywhere on the unit circle $\mathbb{T}$ outside an exceptional set of outer logarithmic capacity zero. Assuming that $f \not\equiv 0$, what can be said about the set $\{\zeta \in \mathbb{T} : f^*(\zeta) = 0\}$? Clearly this set must have Lebesgue measure zero. We shall see at the end of the section that this necessary condition is not sufficient. In fact, no completely satisfactory characterization of such sets appears to be known. In this section and the next, we shall present two important classes of sets that arise as boundary zeros of functions in $\mathcal{D}$: sets of capacity zero and so-called Carleson sets.

We begin with sets of capacity zero. Before treating the general case, it is instructive to consider the special case of countable sets.

Theorem 4.3.1 *Let E be a countable subset of $\mathbb{T}$. Then there exists a function $f \in \mathcal{D} \cap H^\infty$ such that $f(z) \neq 0$ for all $z \in \mathbb{D}$ and $\lim_{r\to 1} f(r\zeta) = 0$ for all $\zeta \in E$.*

Proof Let us write $E = \{\zeta_1, \zeta_2, \dots\}$. For each $n \geq 1$, let Ω_n be the domain defined by

$$\Omega_n := \{x + iy : x > 0,\ |y| < 1/(n^4(1+x)^2)\}.$$

Let $F_n : \mathbb{D} \to \Omega_n$ be the Riemann mapping, normalized so that $F_n(0) = 1/n^2$ and $\lim_{r\to 1} \operatorname{Re} F_n(r\zeta_n) = \infty$. Note that

$$\mathcal{D}(F_n) = \operatorname{area}(\Omega_n)/\pi = 2/(\pi n^4) \qquad (n \geq 1),$$

and so

$$\|F_n\|_{\mathcal{D}}^2 \leq |F_n(0)|^2 + 2\mathcal{D}(F_n) \leq (1 + 4/\pi)/n^4 \qquad (n \geq 1).$$

Consequently $\sum_n \|F_n\|_{\mathcal{D}} < \infty$, and the series $\sum_{n\geq 1} F_n$ converges in $\mathcal{D}$ to a function F. Since $\operatorname{Re} F_n \geq 0$ for all n, we clearly have $\operatorname{Re} F \geq 0$. Set $f := e^{-F}$. Then f is bounded and nowhere zero on $\mathbb{D}$. Further, since $|f'| = |F'e^{-F}| \leq |F'|$, we have $f \in \mathcal{D}$. Finally, for each $\zeta_n \in E$,

$$\lim_{r\to 1} |f(r\zeta_n)| = \lim_{r\to 1} e^{-\operatorname{Re} F(r\zeta_n)} \leq \lim_{r\to 1} e^{-\operatorname{Re} F_n(r\zeta_n)} = 0.$$

Thus f has all the required properties. □

The preceding result applies to arbitrary countable subsets of $\mathbb{T}$. If we restrict ourselves just to closed sets, then we can also treat uncountable sets of capacity zero. The argument is basically similar, the role of the Riemann mappings now being played by certain functions already constructed in the previous chapter. As usual, $c(\cdot)$ denotes logarithmic capacity, as defined in Chapter 2. Also, we write $A(\mathbb{D})$ for the *disk algebra*, namely the algebra of continuous functions $f : \overline{\mathbb{D}} \to \mathbb{C}$ that are holomorphic in $\mathbb{D}$.

Theorem 4.3.2 *Let E be a closed subset of $\mathbb{T}$ with $c(E) = 0$. Then there exists $f \in \mathcal{D} \cap A(\mathbb{D})$ such that $E = \{z \in \overline{\mathbb{D}} : f(z) = 0\}$.*

Proof According to Theorem 3.4.1, there exists a function F continuous on $\overline{\mathbb{D}} \setminus E$ and holomorphic in $\mathbb{D}$ such that $F \in \mathcal{D}$ and $\lim_{z\to\zeta} \operatorname{Re} F(z) = \infty$ for all $\zeta \in E$. Set $f := e^{-F}$. Then $f \in A(\mathbb{D})$ and $f^{-1}(\{0\}) = E$. Also, since $f' = -fF'$, we have $\mathcal{D}(f) \leq \|f\|_\infty^2 \mathcal{D}(F) < \infty$, so $f \in \mathcal{D}$. □

Thus 'capacity zero' is a sufficient condition for a closed set to be a boundary zero set. We shall see in the next section that it is not necessary. However, we do have the following partial converse to Theorem 4.3.2.

Theorem 4.3.3 *Let E be a closed subset of $\mathbb{T}$ with $c(E) > 0$. Then there exists a positive sequence (λ_n) converging to 0 such that $\cup_{n\geq 1}(e^{i\lambda_n}E)$ is a boundary uniqueness set in the following sense: if $f \in \mathcal{D} \cap H^\infty$ and $f^* = 0$ q.e. on $\cup_{n\geq 1}(e^{i\lambda_n}E)$, then $f \equiv 0$.*

Proof Since $c(E) > 0$, there exists a probability measure μ on E whose energy $I_K(\mu)$ is finite. By Theorem 2.4.4, we have $\sum_{k\geq 1} |\widehat{\mu}(k)|^2/k < \infty$. Let (ρ_k) be a positive sequence such that $\rho_k \to \infty$ but still $\sum_k \rho_k^2 |\widehat{\mu}(k)|^2/k < \infty$. For $t > 0$, define

$$\omega(t) := \sup_{k\geq 1} \frac{\min\{2, kt\}}{\rho_k}.$$

Since $\rho_k \to \infty$, it follows that $\omega(t) \to 0$ as $t \to 0^+$. We may thus choose a sequence $(l_n)_{n\geq 1}$ satisfying the conditions $\sum_n l_n = 1$ and $\sum_n l_n \log \omega(l_n) = -\infty$ (see Exercise 4.3.1). Finally, we set $\lambda_n := \sum_{k\geq n} l_k$. We shall prove that, with this choice of (λ_n), the set $\cup_{n\geq 1}(e^{i\lambda_n}E)$ is a uniqueness set for $\mathcal{D} \cap H^\infty$.

Let $f \in \mathcal{D} \cap H^\infty$ with $f^* = 0$ q.e. on $\cup_{n\geq 1}(e^{i\lambda_n}E)$. Define $g : \mathbb{D} \to \mathbb{C}$ by

$$g(z) := \int_E f(z\zeta)\, d\mu(\zeta) \qquad (z \in \mathbb{D}).$$

Clearly $g \in H^\infty$, and by the dominated convergence theorem,

$$g^*(e^{i\lambda_n}) = \int_E f^*(e^{i\lambda_n}\zeta)\, d\mu(\zeta) = 0 \qquad (n \geq 1),$$

the last equality because f^* vanishes q.e. (and hence μ-a.e.) on $e^{i\lambda_n}E$. Now the Taylor coefficients of g are related to those of f by

$$\widehat{g}(k) = \widehat{f}(k)\widehat{\mu}(-k) = \widehat{f}(k)\overline{\widehat{\mu}(k)} \qquad (k \geq 0).$$

By the Cauchy–Schwarz inequality, it follows that

$$\sum_{k\geq 1} |\widehat{g}(k)|\rho_k \leq \Big(\sum_{k\geq 1} k|\widehat{f}(k)|^2|\Big)^{1/2}\Big(\sum_{k\geq 1} \frac{|\widehat{\mu}(k)|^2}{k}\rho_k^2\Big)^{1/2} < \infty.$$

Hence g is continuous on $\overline{\mathbb{D}}$, and satisfies

$$|g(e^{i\theta_1}) - g(e^{i\theta_2})| \leq \sum_{k\geq 1} |\widehat{g}(k)|\rho_k \frac{|2\sin k(\theta_1 - \theta_2)/2|}{\rho_k} \leq C\omega(|\theta_1 - \theta_2|),$$

where $C := \sum_{k\geq 1} |\widehat{g}(k)|\rho_k$. In particular, since $g(e^{i\lambda_n}) = 0$, we have

$$|g(e^{i\theta})| \leq C\omega(l_n) \qquad (\lambda_{n+1} < \theta < \lambda_n),$$

and hence

$$\int_0^1 \log|g(e^{i\theta})|\, d\theta = \sum_{n\geq 1} \int_{\lambda_{n+1}}^{\lambda_n} \log|g(e^{i\theta})|\, d\theta \leq \log C + \sum_{n\geq 1} l_n \log\omega(l_n) = -\infty.$$

As g is bounded, this implies that $g \equiv 0$, and in particular $f(0) = g(0) = 0$. If we repeat the argument successively with $f(z)/z^k$, $k = 0, 1, 2, \ldots$, we find that $f^{(k)}(0) = 0$ for all k. Hence $f \equiv 0$. □

Corollary 4.3.4 *There exists a closed subset F of $\mathbb{T}$ of measure zero such that, if $f \in \mathcal{D} \cap H^\infty$ and $f^* = 0$ q.e. on F, then $f \equiv 0$.*

Proof Let E be the circular Cantor middle-third set. Then $c(E) > 0$, so there exists a sequence (λ_n) tending to zero such that $\cup_n(e^{i\lambda_n}E)$ is a uniqueness set in the sense of the theorem. Set $F := \cup_n(e^{i\lambda_n}E) \cup E$. Then F is a closed set of measure zero and it too is a uniqueness set for $\mathcal{D} \cap H^\infty$. □

In fact, though the boundedness of f was needed in the proof of Theorem 4.3.3, both the theorem and its corollary remain true without it. One way to see this is to show that every function in $\mathcal{D}$ can be expressed as the quotient of two bounded functions in $\mathcal{D}$ (see Exercise 7.5.1).

Exercises 4.3

1. Prove the following fact, needed in the proof of Theorem 4.3.3. Given a function $\omega : (0,\infty) \to (0,\infty)$ such that $\lim_{t\to 0^+} \omega(t) = 0$, there exists a positive sequence (l_n) such that $\sum_n l_n = 1$ and $\sum_n l_n \log \omega(l_n) = -\infty$.

4.4 Boundary zeros II: Carleson sets

Given a closed subset E of $\mathbb{T}$ of logarithmic capacity zero, Theorem 4.3.2 yields a function $f \in \mathcal{D} \cap A(\mathbb{D})$ having E as its zero set. Can the degree of smoothness of f be further improved? For example, can f be chosen so that f' is continuous up to the boundary?

It turns out that a new phenomenon intervenes. For $n \geq 1$, let us write $A^n(\mathbb{D}) := \{f \in A(\mathbb{D}) : f^{(n)} \in A(\mathbb{D})\}$. Note that $A^{n+1}(\mathbb{D}) \subset A^n(\mathbb{D})$ for all $n \geq 1$, and that $A^1(\mathbb{D}) \subset \mathcal{D} \cap A(\mathbb{D})$. It is well known that the boundary zero sets of functions in $A(\mathbb{D})$ are precisely the closed sets of Lebesgue measure zero. For $n \geq 1$, however, the zero sets of $A^n(\mathbb{D})$ are more restricted. Beurling [16] discovered the following necessary condition, and Carleson [26] proved that it is also sufficient.

Definition 4.4.1 A *Carleson set* is a closed subset E of $\mathbb{T}$ such that

$$\int_{\mathbb{T}} \log\Big(\frac{1}{\operatorname{dist}(\zeta, E)}\Big)\,|d\zeta| < \infty. \tag{4.8}$$

The following theorem establishes the necessity of this condition.

Theorem 4.4.2 *Let $f \in A^1(\mathbb{D})$ with $f \not\equiv 0$, and let $E := \{\zeta \in \mathbb{T} : f(\zeta) = 0\}$. Then E is a Carleson set.*

Proof Let $M := \sup_{\overline{\mathbb{D}}} |f'|$. Then $|f(z) - f(w)| \leq M|z - w|$ for all $z, w \in \overline{\mathbb{D}}$. In particular, $|f(z)| \leq M \operatorname{dist}(z, E)$ for all $z \in \overline{\mathbb{D}}$. It follows that

$$\int_{\mathbb{T}} \log|f(\zeta)|\,|d\zeta| \leq 2\pi \log M + \int_{\mathbb{T}} \log(\operatorname{dist}(\zeta, E))\,|d\zeta|.$$

On the other hand, since $f \in H^2 \setminus \{0\}$, we certainly have

$$\int_{\mathbb{T}} \log|f(\zeta)|\,|d\zeta| > -\infty.$$

Therefore (4.8) holds. □

Now we turn to sufficiency.

Theorem 4.4.3 *Let E be a Carleson set and let $n \geq 1$. Then there exists an outer function $f \in A^n(\mathbb{D})$ such that $E = \{\zeta \in \mathbb{T} : f(\zeta) = 0\}$. Furthermore, f may be chosen so that $f^{(k)} \equiv 0$ on E for $k = 0, \dots, n$ and so that f can be continued holomorphically across every point of $\mathbb{T} \setminus E$.*

In particular, Carleson sets are zero sets of functions in $\mathcal{D}$, and these functions can be chosen to be C^n-smooth up to the boundary. Note that the conditions of being a Carleson set and of having capacity zero are independent (see Exercise 4.4.4), so neither one of the Theorems 4.3.2 and 4.4.3 implies the other.

Proof Let $(I_j)_{j\geq 1}$ be the connected components of $\mathbb{T} \setminus E$ and, for each j, let α_j, β_j be the endpoints of I_j. Define $\phi : \mathbb{T} \to \mathbb{R}^+$ by

$$\phi(\zeta) := \begin{cases} |(\zeta - \alpha_j)(\zeta - \beta_j)|, & \zeta \in I_j, \\ 0, & \zeta \in E. \end{cases}$$

Note that $\operatorname{dist}(\zeta, E)^2 \leq \phi(\zeta) \leq 2 \operatorname{dist}(\zeta, E)$, so by (4.8) $\log \phi \in L^1(\mathbb{T})$. Let $N \geq 1$ and let f be the outer function such that $|f^*| = \phi^N$ a.e. on $\mathbb{T}$. Explicitly,

$$f(z) := \exp\Big(\frac{N}{2\pi} \int_{\mathbb{T}} \frac{\zeta + z}{\zeta - z} \log \phi(\zeta)\, |d\zeta|\Big) \qquad (z \in \mathbb{D}).$$

We shall prove that this f works provided that $N > 2n$.

We first show that f extends holomorphically across every point of $\mathbb{T} \setminus E$. For each j, the function $(1 - z/\alpha_j)(1 - z/\beta_j)$ is outer, so we can write it as

$$(1 - z/\alpha_j)(1 - z/\beta_j) = \exp\Big(\frac{1}{2\pi} \int_{\mathbb{T}} \frac{\zeta + z}{\zeta - z} \log |(\zeta - \alpha_j)(\zeta - \beta_j)|\, |d\zeta|\Big) \quad (z \in \mathbb{D}).$$

Dividing this into the formula for f, we obtain the decomposition

$$f(z) = (1 - z/\alpha_j)^N (1 - z/\beta_j)^N e^{N h_j(z)} \qquad (z \in \mathbb{D}), \tag{4.9}$$

where

$$h_j(z) := \frac{1}{2\pi} \int_{\mathbb{T} \setminus I_j} \frac{\zeta + z}{\zeta - z} \log\Big(\frac{\phi(\zeta)}{|(\zeta - \alpha_j)(\zeta - \beta_j)|}\Big) |d\zeta|.$$

Notice that the integral defining h_j can be taken over $\mathbb{T} \setminus I_j$, because the integrand vanishes on I_j. This shows that h_j extends holomorphically across the arc I_j, and therefore, by (4.9), so too does f. Repeating this for every I_j, we deduce that f extends holomorphically across each point of $\mathbb{T} \setminus E$, as claimed.

Next, we establish the following estimate for f:

$$|f(z)| \leq 2^N \operatorname{dist}(z, E)^N \qquad (z \in \mathbb{D}). \tag{4.10}$$

Let $w \in E$. For each $\zeta \in \mathbb{T}$ we have $\phi(\zeta) \leq 2|\zeta - w|$, and so

$$|f(z)| \leq \exp\Big(\frac{N}{2\pi}\int_{\mathbb{T}} \operatorname{Re}\Big(\frac{\zeta + z}{\zeta - z}\Big)\log(2|\zeta - w|)\,|d\zeta|\Big) = 2^N|z - w|^N \qquad (z \in \mathbb{D}).$$

As this holds for all $w \in E$, we obtain (4.10). In particular, f extends continuously to $\overline{\mathbb{D}}$ and $f = 0$ on E.

Lastly, we seek analogous estimates for $f^{(k)}$. For this, we re-use the decomposition (4.9). Differentiating k times the formula for h_j gives

$$h_j^{(k)}(z) = \frac{1}{2\pi}\int_{\mathbb{T}\setminus I_j} \frac{(2\zeta)k!}{(\zeta - z)^{k+1}} \log\Big(\frac{\phi(\zeta)}{|(\zeta - \alpha_j)(\zeta - \beta_j)|}\Big)\,|d\zeta|,$$

and consequently

$$|h_j^{(k)}(z)| \leq \frac{Ck!}{\operatorname{dist}(z, \mathbb{T}\setminus I_j)^{k+1}} \qquad (z \in \mathbb{D}),$$

where C is a constant independent of j, k. Applying Leibniz' formula to (4.9), we deduce that

$$|f^{(k)}(z)| \leq \frac{C_k|f(z)|}{\operatorname{dist}(z, \mathbb{T}\setminus I_j)^{2k}} \qquad (z \in \mathbb{D}),$$

where the C_k are constants depending on k, but not on j. Repeating for each I_j, we obtain

$$|f^{(k)}(z)| \leq \frac{C_k|f(z)|}{\operatorname{dist}(z, E)^{2k}} \qquad (z \in \mathbb{D}).$$

In combination with the estimate (4.10), this yields

$$|f^{(k)}(z)| \leq 2^N C_k \operatorname{dist}(z, E)^{N-2k} \qquad (z \in \mathbb{D}).$$

Thus, provided that $N > 2n$, the functions $f, f', \ldots, f^{(n)}$ all extend continuously to $\overline{\mathbb{D}}$ and are equal to zero on E. □

Exercises 4.4

1. (i) Show that a closed subset of a Carleson set is again a Carleson set.
   (ii) Show that the union of two Carleson sets is again a Carleson set.
2. Let E be a closed subset of $\mathbb{T}$, and let (I_k) be the components of $\mathbb{T}\setminus E$. Show that E is a Carleson set if and only if $|E| = 0$ and $\sum_k |I_k|\log(1/|I_k|) < \infty$.
3. Which of the following subsets of $\mathbb{T}$ are Carleson sets?
   (i) $\{e^{i/n} : n \geq 1\} \cup \{1\}$.
   (ii) $\{e^{i/\log n} : n \geq 2\} \cup \{1\}$.
   (iii) $\{e^{i\theta} : \theta \in C\}$, where C is the Cantor middle-third set.

4. Give examples of closed subsets E of $\mathbb{T}$ to show that neither one of the conditions 'E is a Carleson set' and 'E is of logarithmic capacity zero' implies the other. [Hint: Use the preceding exercise.]

4.5 Arguments of zero sets

We now return to the problem of describing zero sets of $\mathcal{D}$ in the interior of the unit disk. Let us recall what we have seen so far. Given a sequence (r_n) in $(0, 1)$, there are three possibilities.

- If $\sum_n(1 - r_n) = \infty$, then every sequence (z_n) with $|z_n| = r_n$ is a uniqueness set for $\mathcal{D}$ (because it is true even for H^2).
- If $\sum_n 1/\log(1/(1 - r_n)) < \infty$, then every sequence (z_n) with $|z_n| = r_n$ is a zero set for $\mathcal{D}$ (Theorem 4.2.1).
- If $\sum_n(1 - r_n) < \infty$ and $\sum_n 1/\log(1/(1 - r_n)) = \infty$, then there exist sequences (z_n) and (z'_n) with $|z_n| = |z'_n| = r_n$ for all n, such that (z_n) is a zero set and (z'_n) is a uniqueness set (see Theorems 4.1.2 and 4.2.2.)

Thus, in the third case, the arguments of the (z_n) play a critical role in determining whether (z_n) is a zero set or a uniqueness set. This is the subject of this section.

The following theorem is a substantial generalization of Theorem 4.1.2, and is in some sense dual to Theorem 4.2.1.

Theorem 4.5.1 *Let E be a Carleson set. If (z_n) is a Blaschke sequence such that $z_n/|z_n| \in E$ for all n, then (z_n) is a zero set for $\mathcal{D}$.*

The theorem is sharp in the following sense.

Theorem 4.5.2 *Let $(\theta_n)_{n\geq 1}$ be a sequence such that $\overline{\{e^{i\theta_n} : n \geq 1\}}$ is not a Carleson set. Then there exists a positive Blaschke sequence (r_n) such that $(r_n e^{i\theta_n})$ is a uniqueness set for $\mathcal{D}$.*

We now turn to the proofs of the two theorems.

Proof of Theorem 4.5.1 By Theorem 4.4.3, since E is a Carleson set, there exists $f \in A^1(\mathbb{D})$ such that $f = 0$ on E. In particular, we have $f \in \mathcal{D}$ and $|f(z)| \leq C\,\mathrm{dist}(z, E)$ for some constant C. Given a Blaschke sequence (z_n), let B be the Blaschke product with zero set (z_n). Clearly $(Bf)(z_n) = 0$ for all n, and we shall now show that $Bf \in \mathcal{D}$.

By Theorem 4.1.3, we have

$$\mathcal{D}(Bf) = \mathcal{D}(f) + \sum_n \frac{1}{2\pi} \int_{\mathbb{T}} \frac{1 - |z_n|^2}{|\zeta - z_n|^2} |f(\zeta)|^2 \, |d\zeta|.$$

Since $f \in \mathcal{D}$ we certainly have $\mathcal{D}(f) < \infty$. Also, since $|f(z)| \leq C \operatorname{dist}(z, E)$ and $z_n/|z_n| \in E$, we have $|f(\zeta)| \leq C|\zeta - z_n/|z_n|| \leq 2C|\zeta - z_n|$ for all n, where the last inequality comes from Lemma 1.6.7. Consequently,

$$\sum_n \frac{1}{2\pi} \int_{\mathbb{T}} \frac{1 - |z_n|^2}{|\zeta - z_n|^2} |f(\zeta)|^2 \, |d\zeta| \leq 4C^2 \sum_n (1 - |z_n|^2) < \infty.$$

Thus $\mathcal{D}(Bf) < \infty$, and the proof is complete. □

For the proof of Theorem 4.5.2, we need a lemma.

Lemma 4.5.3 *Let E be a subset of $\mathbb{T}$ whose closure is not a Carleson set. Then there exists a sequence in E which converges and whose closure is not a Carleson set.*

Proof By assumption, we have

$$\int_{\mathbb{T}} \log\Bigl(\frac{1}{\operatorname{dist}(\zeta, E)}\Bigr) |d\zeta| = \infty.$$

By the compactness of $\mathbb{T}$, there exists $\zeta_0 \in \mathbb{T}$ such that, for every neighborhood I of ζ_0,

$$\int_I \log\Bigl(\frac{1}{\operatorname{dist}(\zeta, E)}\Bigr) |d\zeta| = \infty.$$

Let (I_n) be a decreasing sequence of neighborhoods of ζ_0 whose intersection is equal to $\{\zeta_0\}$. For each n, we may pick a finite subset F_n of $E \cap I_n$ such that

$$\int_{I_n} \log\Bigl(\frac{1}{\operatorname{dist}(\zeta, F_n)}\Bigr) |d\zeta| \geq n.$$

Let $S := \cup_n F_n$. Then $S \subset E$ and S is a sequence converging to ζ_0. Finally, S contains F_n for each n, so

$$\int_{\mathbb{T}} \log\Bigl(\frac{1}{\operatorname{dist}(\zeta, S)}\Bigr) |d\zeta| \geq \sup_n \int_{\mathbb{T}} \log\Bigl(\frac{1}{\operatorname{dist}(\zeta, F_n)}\Bigr) |d\zeta| = \infty,$$

and thus the closure of S is not a Carleson set. □

Proof of Theorem 4.5.2 As $\overline{\{e^{i\theta_n} : n \geq 1\}}$ is not a Carleson set, Lemma 4.5.3 guarantees that there is a convergent subsequence of the $(e^{i\theta_n})$ whose closure is still not a Carleson set. Relabeling, we may as well suppose that this is the whole sequence. Furthermore, since the union of two Carleson sets is again a

Carleson set, we can suppose that $(e^{i\theta_n})$ converges to its limit from one side. Relabeling again, we may thus assume that (θ_n) is either increasing or decreasing. We shall suppose that it is decreasing, the proof for the increasing case being practically the same, with obvious modifications. Finally, there is no loss of generality in supposing that $0 < \theta_{n+1} < \theta_n < 1$ for all n and that $\theta_n \to 0$.

For $n \geq 1$, set $\delta_n := \theta_n - \theta_{n+1}$. Clearly $\sum_n \delta_n < \infty$. Also, since the closure of $\{e^{i\theta_n} : n \geq 1\}$ is not a Carleson set, we have $\sum_n \delta_n \log(1/\delta_n) = \infty$. We are now in the situation of Theorem 4.1.6. By that result, if we set $r_n := 1-\delta_n$, then $(r_n e^{i\theta_n})$ is a uniqueness set for $\mathcal{D}$, and obviously (r_n) is a Blaschke sequence. □

Exercises 4.5

1. Show that there exist zero sets for $\mathcal{D}$ of the form $((1 - n^{-2})e^{i\theta_n})$ for which $\theta_n \to 0$ arbitrarily slowly.

Notes on Chapter 4

§4.1

Theorems 4.1.2 and 4.1.3 are due to Carleson [25]. Theorems 4.1.5 and 4.1.6 are due to Caughran [31].

§4.2

Theorem 4.2.1 is due to H. Shapiro and Shields [110], and Theorem 4.2.2 is due to Nagel, Rudin and J. Shapiro [83]. Slightly weaker versions of both results had earlier been obtained by Carleson in [25]. A different proof of Theorem 4.2.2, due to Richter, Ross and Sundberg, can be found in [101]. Their construction yields a uniqueness set that accumulates at just one point of the unit circle. Lemma 4.2.7 is a result of Kaluza (see [55, p. 68 and p. 91]). Exercise 4.2.3 is taken from [63], where it is also shown that there exist sequences (z_n) that are zero sets for $\mathcal{D}$, but for which $\mathcal{E}$ is of positive logarithmic capacity.

§4.3

Theorem 4.3.1 is due to Rudin [105]. Theorem 4.3.2 is a result of Brown and Cohn [23], improving an earlier theorem of Carleson [26, Theorem 4]. In fact, rather more is true: given a closed set E of capacity zero and given a continuous function h on E, there exists $f \in \mathcal{D} \cap A(\mathbb{D})$ such that $f|_E = h$. This interpolation theorem is due to Peller and Hruščev [90]; another proof, due to Koosis, can be found in [123, IIIE.6]. Theorem 4.3.3 is due to Carleson [25].

§4.4

Carleson sets are also referred to in the literature as Beurling–Carleson sets and even as Beurling–Carleson–Hayman sets. An account of their history can be found in Shapiro's article [111], which also contains a lot of further information about them. Theorem 4.4.2 is due to Beurling [16] and Theorem 4.4.3 is due to Carleson [26]. Carleson's theorem can be extended to yield a function $f \in A^\infty(\mathbb{D}) := \cap_{n\geq 1}A^n(\mathbb{D})$ with zero set E: see [69, 86, 118].

In [26], Carleson also proved the following result in the other direction: if E is a closed subset of $\mathbb{T}$ which is not a Carleson set, and if, in addition, for some $\alpha \in (0,1)$, the Riesz capacity c_α satisfies

$$c_\alpha(E \cap I) \geq \text{const.}|I|$$

for all arcs I with centers belonging to E, then E is a uniqueness set for $\mathcal{D}$. Further extensions of this result were obtained by Maz′ya and Havin [79]; see also [62]. We also mention the article of Malliavin [74], in which the author characterizes boundary uniqueness sets for $\mathcal{D}$ in terms of a modified capacity, defined using a complex-valued kernel.

§4.5

Theorems 4.5.1 and 4.5.2 are due to Caughran [32], though the proofs here are a bit different. The article [63] contains generalizations of Theorem 4.5.1 where, instead of stipulating that the arguments of the z_n belong to E, one assumes merely that they converge to E suitably fast. Another result along these lines can be found in the article of Bogdan [19]. The proof of Lemma 4.5.3 is taken from [68]; a further proof can be found in [78].

5

Multipliers

The Dirichlet space $\mathcal{D}$ is not an algebra. We saw this in §1.3, where we also proved that $\mathcal{D} \cap H^\infty$ *is* an algebra, even a Banach algebra with respect to a suitable norm. This is reminiscent of the case of Hardy spaces: H^2 is not an algebra, but H^∞ is one. However, there is also a significant difference between the Dirichlet and Hardy cases. The algebra H^∞ derives much of its importance from the fact that its elements are multipliers of H^2, in other words, that

$$h \in H^\infty, f \in H^2 \Rightarrow hf \in H^2.$$

On the other hand, it turns out that *not* all elements of $\mathcal{D} \cap H^\infty$ are multipliers of $\mathcal{D}$. The functions that are multipliers of $\mathcal{D}$ form an algebra in their own right which, for many purposes, is the correct analogue of H^∞ in the Dirichlet-space setting.

In this chapter we shall study the multipliers of $\mathcal{D}$, attempting to characterize them, and establishing Dirichlet-space analogues of a number of basic results about H^∞ concerning zeros and interpolation.

5.1 Definition and elementary properties

Definition 5.1.1 A *multiplier* of $\mathcal{D}$ is a function $h : \mathbb{D} \to \mathbb{C}$ such that $hf \in \mathcal{D}$ whenever $f \in \mathcal{D}$. The set of multipliers is an algebra, called the *multiplier algebra* of $\mathcal{D}$. We denote it by $\mathcal{M}(\mathcal{D})$.

Clearly every multiplier of $\mathcal{D}$ is itself a member of $\mathcal{D}$. Indeed, if $h \in \mathcal{M}(\mathcal{D})$, then $h = h1 \in \mathcal{D}$. In the other direction, every polynomial is a multiplier of $\mathcal{D}$. More generally, if $h \in \operatorname{Hol}(\mathbb{D})$ and h' is bounded on $\mathbb{D}$, then h is a multiplier. To see this, note that, given $f \in \mathcal{D}$, we have $(hf)' = h'f+hf' = h'(zf)'+(h-zh')f'$, and so

$$\mathcal{D}(hf) \le 2\|h'\|_{H^\infty}^2 \mathcal{D}(zf) + 2\|h - zh'\|_{H^\infty}^2 \mathcal{D}(f) < \infty.$$

Further examples of multipliers are given in Exercise 5.1.3.

Given $h \in \mathcal{M}(\mathcal{D})$, we write $M_h : \mathcal{D} \to \mathcal{D}$ for the multiplication operator

$$M_h(f) := hf \qquad (f \in \mathcal{D}).$$

Theorem 5.1.2 *If $h \in \mathcal{M}(\mathcal{D})$, then M_h is a continuous linear map from $\mathcal{D}$ to itself.*

Proof We use the closed graph theorem. Suppose that $f_n \to 0$ in $\mathcal{D}$ and that $M_h(f_n) \to g$ in $\mathcal{D}$. We need to show that $g = 0$. For each $z \in \mathbb{D}$, evaluation at z is a continuous linear functional on $\mathcal{D}$, so $g(z) = \lim_n M_h(f_n)(z) = h(z) \lim_n f_n(z) = 0$. Hence $g = 0$, and we are done. □

This theorem allows us to define the *multiplier norm* $\|\cdot\|_{\mathcal{M}(\mathcal{D})}$ on $\mathcal{M}(\mathcal{D})$ by

$$\|h\|_{\mathcal{M}(\mathcal{D})} := \|M_h\| = \sup\{\|hf\|_{\mathcal{D}} : \|f\|_{\mathcal{D}} \le 1\} \qquad (h \in \mathcal{M}(\mathcal{D})).$$

Taking $f = 1$, we see that $\|h\|_{\mathcal{M}(\mathcal{D})} \ge \|h\|_{\mathcal{D}}$ for all $h \in \mathcal{M}(\mathcal{D})$.

Theorem 5.1.3 *$(\mathcal{M}(\mathcal{D}), \|\cdot\|_{\mathcal{M}(\mathcal{D})})$ is a Banach algebra.*

Proof It is obviously a normed algebra; all that is left to prove is completeness. Suppose that (h_n) is a Cauchy sequence in $\mathcal{M}(\mathcal{D})$. Then (M_{h_n}) is a Cauchy sequence in the Banach algebra of $B(\mathcal{D})$ of bounded linear operators on $\mathcal{D}$, so there exists $T \in B(\mathcal{D})$ such that $\|M_{h_n} - T\|_{B(\mathcal{D})} \to 0$. Thus $\|h_n f - T(f)\|_{\mathcal{D}} \to 0$ for each $f \in \mathcal{D}$. In particular $\|h_n - T(1)\|_{\mathcal{D}} \to 0$. So, if we set $h := T(1)$, then $h \in \mathcal{D}$ and $h_n \to h$ pointwise on $\mathbb{D}$. It follows that $T(f) = hf$ for all $f \in \mathcal{D}$. Hence $h \in \mathcal{M}(\mathcal{D})$, and $\|h_n - h\|_{\mathcal{M}(\mathcal{D})} = \|M_{h_n} - T\|_{B(\mathcal{D})} \to 0$ as $n \to \infty$. □

The next result establishes an important connection between multipliers and reproducing kernels. We recall from §1.2 that, if

$$k_w(z) := \frac{1}{z\overline{w}} \log\Big(\frac{1}{1 - z\overline{w}}\Big) \qquad (z \in \mathbb{D}),$$

then $k_w \in \mathcal{D}$ and

$$f(w) = \langle f, k_w \rangle_{\mathcal{D}} \qquad (w \in \mathbb{D}).$$

Also, given an operator T on a Hilbert space, we write T^* for its adjoint.

Theorem 5.1.4 *If $h \in \mathcal{M}(\mathcal{D})$, then*

$$M_h^*(k_w) = \overline{h(w)} k_w \qquad (w \in \mathbb{D}).$$

Conversely, given a bounded operator $T : \mathcal{D} \to \mathcal{D}$ such that the k_w are eigenvectors of T^*, there exists $h \in \mathcal{M}(\mathcal{D})$ such that $T = M_h$.

Proof Let $w \in \mathbb{D}$. If $f \in \mathcal{D}$, then

$$\langle f, M_h^*(k_w)\rangle_{\mathcal{D}} = \langle M_h(f), k_w\rangle_{\mathcal{D}} = \langle hf, k_w\rangle_{\mathcal{D}} = h(w)f(w) = \langle f, \overline{h(w)}k_w\rangle_{\mathcal{D}}.$$

As this holds for all $f \in \mathcal{D}$, we deduce that $M_h^*(k_w) = \overline{h(w)}k_w$.

For the converse, let $h(w)$ be the complex conjugate of the eigenvalue of T^* corresponding to the eigenvector k_w. Then, for each $f \in \mathcal{D}$, we have

$$(Tf)(w) = \langle Tf, k_w\rangle_{\mathcal{D}} = \langle f, T^*k_w\rangle_{\mathcal{D}} = h(w)\langle f, k_w\rangle_{\mathcal{D}} = h(w)f(w).$$

Hence h is a multiplier and $T = M_h$. □

This result, though simple, has many useful consequences. One is given in Exercise 5.1.4, and another will appear when we come to study interpolation in §5.3. Here, we shall use it to establish an important property of multipliers.

Theorem 5.1.5 *If $h \in \mathcal{M}(\mathcal{D})$, then h is bounded and $\|h\|_{H^\infty} \leq \|h\|_{\mathcal{M}(\mathcal{D})}$.*

Proof Let $h \in \mathcal{M}(\mathcal{D})$. Then M_h^* is a bounded operator on $\mathcal{D}$, so its eigenvalues are bounded, and of modulus no more than $\|M_h\|$. By Theorem 5.1.4, it follows that the values of h are bounded, and of modulus no more than $\|h\|_{\mathcal{M}(\mathcal{D})}$. □

We have now seen that $\mathcal{M}(\mathcal{D}) \subset \mathcal{D} \cap H^\infty$. Also, we showed in Theorem 1.3.2 that $\mathcal{D} \cap H^\infty$ is a Banach algebra. So it is very tempting to guess that this inclusion is in fact an equality, all the more so since the analogous result for the Hardy space (namely $\mathcal{M}(H^2) = H^\infty$) is actually true. However, for the Dirichlet space, this guess turns out to be false, as we now show.

Theorem 5.1.6 *There exists $h \in \mathcal{D} \cap H^\infty$ such that $h \notin \mathcal{M}(\mathcal{D})$.*

Proof For $k \geq 0$, let $n_k := 2^{2^k}$. For $k \geq 1$ and $n_{k-1} \leq n < n_k$, let $a_n := 1/(n_k k^2)$. We shall show that $h(z) := \sum_{n\geq 2} a_n z^n$ satisfies the conclusions of the theorem.

First, we note that

$$\sum_{n\geq 2} a_n = \sum_{k\geq 1}\Big(\sum_{n_{k-1}\leq n<n_k} a_n\Big) = \sum_{k\geq 1}(n_k - n_{k-1})\frac{1}{n_k k^2} \leq \sum_{k\geq 1}\frac{1}{k^2} < \infty,$$

which implies that $h \in H^\infty$ (and even that $h \in A(\mathbb{D})$). Also, we have

$$\sum_{n\geq 2} n a_n^2 = \sum_{k\geq 1}\Big(\sum_{n_{k-1}\leq n<n_k} n a_n^2\Big) \leq \sum_{k\geq 1}(n_k - n_{k-1})\frac{n_k}{n_k^2 k^4} \leq \sum_{k\geq 1}\frac{1}{k^4} < \infty,$$

so $h \in \mathcal{D}$.

It remains to show that h is not a multiplier of $\mathcal{D}$. For this, we consider the function $f(z) := \sum_{n\geq 2} b_n z^n$ where $b_n := 1/(n(\log n)^{3/4})$. Since

$$\sum_{n\geq 2} n b_n^2 = \sum_{n\geq 2}\frac{1}{n(\log n)^{3/2}} < \infty,$$

we have $f \in \mathcal{D}$. On the other hand, $hf \notin \mathcal{D}$. Indeed, $h(z)f(z) = \sum_{n\geq 4} c_n z^n$, where

$$\sum_{n\geq 4} n|c_n|^2 = \sum_{n\geq 4} n\Big(\sum_{j=2}^{n-2} a_{n-j}b_j\Big)^2 \geq \sum_{n\geq 4} na_n^2\Big(\sum_{j=2}^{n-2} b_j\Big)^2.$$

Using the result of Exercise 1.3.1, we have

$$\sum_{j=2}^{n-2} b_j = \sum_{j=2}^{n-2} \frac{1}{j(\log j)^{3/4}} = 4(\log n)^{1/4} + O(1) \qquad (n \to \infty),$$

and elementary estimates show that

$$\begin{aligned}
\sum_{n\geq 4} na_n^2(\log n)^{1/2} &\geq \sum_{k\geq 2}\Big(\sum_{n_k/2\leq n<n_k} na_n^2(\log n)^{1/2}\Big)\\
&\geq \sum_{k\geq 2} \frac{n_k}{2}\frac{(n_k/2)}{n_k^2 k^4}(\log n_{k-1})^{1/2}\\
&= \sum_{k\geq 2} \frac{1}{4k^4}(\log(2^{2^{k-1}}))^{1/2} = \infty.
\end{aligned}$$

Therefore $\sum_n n|c_n|^2 = \infty$, and so $hf \notin \mathcal{D}$, as claimed. We conclude that h is not a multiplier of $\mathcal{D}$. □

Given that membership $\mathcal{D} \cap H^\infty$ does not guarantee being a multiplier, we are led to pose the question of how to characterize the elements of $\mathcal{M}(\mathcal{D})$. The next result is a step towards answering this question.

Theorem 5.1.7 *Let $h \in \mathcal{D}$, and let μ be the finite measure on $\mathbb{D}$ given by $d\mu(z) := |h'(z)|^2\, dA(z)/\pi$. Then $h \in \mathcal{M}(\mathcal{D})$ if and only if h is bounded and there exists a constant C such that*

$$\int_{\mathbb{D}} |f|^2\, d\mu \leq C\|f\|_{\mathcal{D}}^2 \qquad (f \in \mathcal{D}). \tag{5.1}$$

Proof Suppose that $h \in \mathcal{M}(\mathcal{D})$. Then h is bounded by Theorem 5.1.5. Also, given $f \in \mathcal{D}$, we have $fh' = (hf)' - hf'$, and hence

$$\begin{aligned}
\int_{\mathbb{D}} |f|^2\, d\mu &= \frac{1}{\pi}\int_{\mathbb{D}} |fh'|^2\, dA\\
&\leq \frac{1}{\pi}\int_{\mathbb{D}} 2(|(hf)'|^2 + |hf'|^2)\, dA\\
&\leq 2(\|hf\|_{\mathcal{D}}^2 + \|h\|_{H^\infty}^2\|f\|_{\mathcal{D}}^2)\\
&\leq 2(\|h\|_{\mathcal{M}(\mathcal{D})}^2 + \|h\|_{H^\infty}^2)\|f\|_{\mathcal{D}}^2.
\end{aligned}$$

Therefore (5.1) holds, with $C = 2\|h\|^2_{\mathcal{M}(\mathcal{D})} + 2\|h\|^2_{H^\infty}$.

Conversely, suppose that $h \in H^\infty$ and that (5.1) holds. Given $f \in \mathcal{D}$, we have $(hf)' = h'f + hf'$, and hence

$$\begin{aligned}\mathcal{D}(hf) &\le \frac{1}{\pi}\int_{\mathbb{D}} 2(|h'f|^2 + |hf'|^2)\,dA \\ &\le 2\int_{\mathbb{D}} |f|^2\,d\mu + \frac{2}{\pi}\|h\|^2_{H^\infty}\int_{\mathbb{D}} |f'|^2\,dA \\ &\le 2C\|f\|^2_{\mathcal{D}} + 2\|h\|^2_{H^\infty}\|f\|^2_{\mathcal{D}} < \infty.\end{aligned}$$

Therefore h is a multiplier of $\mathcal{D}$, and $\|h\|^2_{\mathcal{M}(\mathcal{D})} \le 2C + 2\|h\|^2_{H^\infty}$. □

This result is not really satisfactory, since it begs the question as to which measures μ on $\mathbb{D}$ satisfy (5.1). This question is the subject of the next section.

Exercises 5.1

1. Show that $\|z^n\|_{\mathcal{M}(\mathcal{D})} = \sqrt{n+1}$ for each $n \ge 0$.
2. Let $(h_n)_{n\ge1} \in \mathcal{M}(\mathcal{D})$. Show that if $h_n \to h$ locally uniformly on $\mathbb{D}$, then $\|h\|_{\mathcal{M}(\mathcal{D})} \le \liminf_{n\to\infty} \|h_n\|_{\mathcal{M}(\mathcal{D})}$. Deduce that, if $\sup_n \|h_n\|_{\mathcal{M}(\mathcal{D})} < \infty$, then $h \in \mathcal{M}(\mathcal{D})$. [Use the result of Exercise 1.1.3.]
3. Let $h \in \operatorname{Hol}(\mathbb{D})$, say $h(z) := \sum_{n\ge0} a_n z^n$.

   (i) Show that
   $$\frac{1}{\pi}\int_{\mathbb{D}} |h'(z)|^2 \log\Big(\frac{1}{1-|z|^2}\Big)\,dA(z) = \sum_{n\ge1} n\Big(\sum_{k=1}^{n}\frac{1}{k}\Big)|a_n|^2.$$

   (ii) Deduce that, if $h \in H^\infty$ and $\sum_n (n\log n)|a_n|^2 < \infty$, then $h \in \mathcal{M}(\mathcal{D})$. [Hint: Use the estimate in Theorem 1.2.1.]

   (iii) Deduce that $\mathcal{D}_\alpha \subset \mathcal{M}(\mathcal{D})$ for all $\alpha \in (-1, 0)$. [Use the results of Exercises 1.6.3 and 1.6.4.]
4. Let $T : \mathcal{D} \to \mathcal{D}$ be a bounded linear map such that $TM_z = M_zT$.

   (i) Show that, for each $w \in \mathbb{D}$, the vector $T^*(k_w)$ is an eigenvector of M_z^* with eigenvalue $\overline{w}$.

   (ii) Show that each eigenspace of M_z^* is one-dimensional, and deduce that each k_w is an eigenvector of T^*.

   (iii) Deduce that there exists $h \in \mathcal{M}(\mathcal{D})$ such that $T = M_h$.

5.2 Carleson measures

Definition 5.2.1 A positive Borel measure μ on $\mathbb{D}$ is called a *Carleson measure* for $\mathcal{D}$ if there exists a constant C such that

$$\int_{\mathbb{D}} |f|^2 \, d\mu \leq C\|f\|_{\mathcal{D}}^2 \qquad (f \in \mathcal{D}). \tag{5.2}$$

Condition (5.2) amounts to saying that $\mathcal{D}$ embeds continuously into $L^2(\mu)$. If $\mathcal{D} \subset L^2(\mu)$, then the embedding is automatically continuous, by the closed graph theorem. Thus in fact μ is a Carleson measure for $\mathcal{D}$ provided simply that $\int_{\mathbb{D}} |f|^2 \, d\mu < \infty$ for all $f \in \mathcal{D}$.

Carleson measures were first introduced by Carleson in his solution of the corona problem. They were originally defined for the Hardy spaces H^p, but the analogous definition makes sense in other function spaces, including the Dirichlet space, and Carleson measures have now become a useful tool in many contexts. For example, Theorem 5.1.7 characterizes the multipliers of $\mathcal{D}$ as being those $h \in H^\infty$ such that $|h'|^2 \, dA$ is a Carleson measure for $\mathcal{D}$. Our aim in this section is to describe Carleson measures for $\mathcal{D}$.

The first result is a dual formulation of the notion of Carleson measure. We recall that k_w denotes the reproducing kernel for $\mathcal{D}$ at w, and we write $k(z, w) := k_w(z)$.

Theorem 5.2.2 *Let μ be a finite positive Borel measure on $\mathbb{D}$. Then*

$$\sup_{\|f\|_{\mathcal{D}} \leq 1} \int_{\mathbb{D}} |f(z)|^2 \, d\mu(z) = \sup_{\|g\|_{L^2(\mu)} \leq 1} \left| \int_{\mathbb{D}} \int_{\mathbb{D}} k(w, z) g(z) \overline{g(w)} \, d\mu(z) \, d\mu(w) \right|. \tag{5.3}$$

Proof Suppose first that the left-hand side of (5.3) is finite, in other words, that μ is a Carleson measure for $\mathcal{D}$. Denote by $J : \mathcal{D} \to L^2(\mu)$ the inclusion map, and by $J^* : L^2(\mu) \to \mathcal{D}$ its adjoint. Clearly the left-hand side of (5.3) is exactly $\|J\|^2$. Also for each $g \in L^2(\mu)$ and each $w \in \mathbb{D}$, we have

$$(J^* g)(w) = \langle J^* g, k_w \rangle_{\mathcal{D}} = \langle g, J k_w \rangle_{L^2(\mu)} = \int_{\mathbb{D}} k(w, z) g(z) \, d\mu(z),$$

so

$$\|J^* g\|_{\mathcal{D}}^2 = \langle J^* g, J^* g \rangle_{\mathcal{D}} = \langle J J^* g, g \rangle_{L^2(\mu)} = \int_{\mathbb{D}} \int_{\mathbb{D}} k(w, z) g(z) \overline{g(w)} \, d\mu(z) \, d\mu(w).$$

Consequently the right-hand side of (5.3) is exactly $\|J^*\|^2$. Thus the fact that (5.3) holds is simply a translation of the standard equality $\|J\| = \|J^*\|$.

Now suppose that the left-hand side of (5.3) is infinite, and let us show that the right-hand side is infinite too. For each $r < 1$, the right-hand side is at least

$$\sup_{\|g\|_{L^2(\mu)}\le 1}\Big|\int_{\mathbb{D}_r}\int_{\mathbb{D}_r} k(w,z)g(z)\overline{g(w)}\,d\mu(z)\,d\mu(w)\Big|,$$

where $\mathbb{D}_r := \{z \in \mathbb{C} : |z| < r\}$. By applying what we have already proved to the measure $\mu|_{\mathbb{D}_r}$ (which certainly is a Carleson measure for $\mathcal{D}$), we see that this last supremum equals

$$\sup_{\|f\|_{\mathcal{D}}\le 1}\int_{\mathbb{D}_r} |f(z)|^2\,d\mu(z).$$

This quantity tends to infinity as $r \to 1$. □

From this result, we deduce an easily applicable sufficient condition for μ to be a Carleson measure for $\mathcal{D}$.

Theorem 5.2.3 *Let μ be a finite positive Borel measure on $\mathbb{D}$ such that*

$$\sup_{w\in\mathbb{D}}\int_{\mathbb{D}} \log\Big|\frac{2}{1-\overline{w}z}\Big|\,d\mu(z) < \infty. \tag{5.4}$$

Then μ is a Carleson measure for $\mathcal{D}$.

Proof By Theorem 5.2.2, it suffices to check that the right-hand side of (5.3) is finite, which we now do. As

$$|k(z,w)| \le 2\log\frac{2}{|1-\overline{w}z|} + \pi \qquad (z,w\in\mathbb{D})$$

(see Exercise 5.2.1), the condition (5.4) implies that

$$M := \sup_{w\in\mathbb{D}}\int_{\mathbb{D}} |k(z,w)|\,d\mu(z) < \infty.$$

By the Cauchy–Schwarz inequality,

$$\begin{aligned}&\Big|\int_{\mathbb{D}}\int_{\mathbb{D}} k(w,z)g(z)\overline{g(w)}\,d\mu(z)\,d\mu(w)\Big|\\ &\le \Big(\int_{\mathbb{D}}\int_{\mathbb{D}} |k(z,w)||g(z)|^2\,d\mu(z)\,d\mu(w)\Big)^{1/2}\Big(\int_{\mathbb{D}}\int_{\mathbb{D}} |k(z,w)||g(w)|^2\,d\mu(z)\,d\mu(w)\Big)^{1/2}.\end{aligned}$$

Since $|k(z,w)| = |k(w,z)|$, we thus have

$$\begin{aligned}\Big|\int_{\mathbb{D}}\int_{\mathbb{D}} k(w,z)g(z)\overline{g(w)}\,d\mu(z)\,d\mu(w)\Big| &\le \int_{\mathbb{D}}\int_{\mathbb{D}} |k(z,w)||g(w)|^2\,d\mu(z)\,d\mu(w)\\ &\le M\|g\|^2_{L^2(\mu)}.\end{aligned}$$

Therefore the right-hand side of (5.3) is indeed finite, as desired. □

Corollary 5.2.4 *Let $h \in H^\infty$ be a function satisfying*

$$\sup_{w\in\mathbb{D}} \int_{\mathbb{D}} |h'(z)|^2 \log\frac{2}{|1-\overline{w}z|}\,dA(z) < \infty.$$

Then h is a multiplier of $\mathcal{D}$.

Proof Combine Theorems 5.1.7 and 5.2.3. □

As mentioned earlier, Carleson measures were originally defined for the Hardy spaces H^p. In this case, they turn out to have a very simple geometric characterization. In order to describe it, we need to introduce some notation.

Given an arc $I \subset \mathbb{T}$, we write

$$S(I) := \{re^{i\theta} : e^{i\theta} \in I,\ 1-|I| < r < 1\}.$$

Here, as usual, $|I|$ denotes the arclength of I. The set $S(I)$ is sometimes called the *Carleson box* corresponding to I.

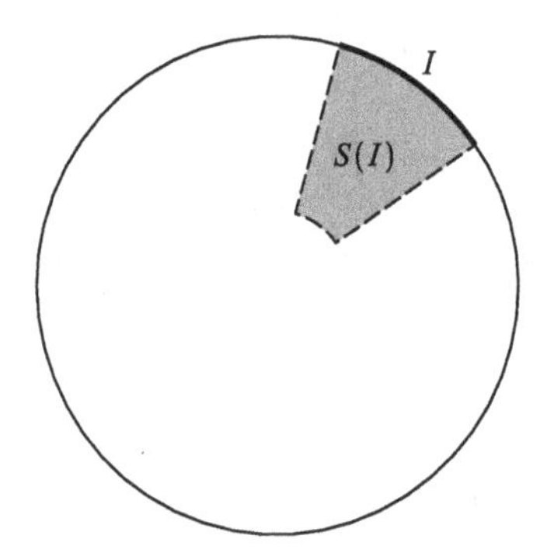

Figure 5.1 A Carleson box

Carleson showed that, if, for some $p \geq 1$,

$$\int_{\mathbb{D}} |f|^2\,d\mu \leq C\|f\|^2_{H^p} \qquad (f \in H^p), \tag{5.5}$$

then, for all arcs I,

$$\mu(S(I)) = O(|I|). \tag{5.6}$$

Conversely, if (5.6) holds, then (5.5) holds for all $p \geq 1$ (even with C independent of p).

We seek a similar characterization of Carleson measures for $\mathcal{D}$. The following result gives separate necessary and sufficient conditions of the same general nature as (5.6).

Theorem 5.2.5 *Let μ be a finite positive Borel measure on $\mathbb{D}$.*

(i) *A necessary condition that μ be a Carleson measure for $\mathcal{D}$ is that*

$$\mu(S(I)) = O\Big(\Big(\log\frac{1}{|I|}\Big)^{-1}\Big) \qquad (|I| \to 0). \tag{5.7}$$

(ii) *A sufficient condition that μ be a Carleson measure for $\mathcal{D}$ is that*

$$\mu(S(I)) = O\big(\phi(|I|)\big) \qquad (|I| \to 0), \tag{5.8}$$

where ϕ is a positive increasing function such that $\int_0^{2\pi}(\phi(x)/x)\,dx < \infty$.

Thus, in particular, the condition that

$$\mu(S(I)) = O\Big(\Big(\log\frac{1}{|I|}\Big)^{-1}\Big(\log\log\frac{1}{|I|}\Big)^{-\alpha}\Big) \qquad (|I| \to 0)$$

is necessary if $\alpha = 0$ and sufficient if $\alpha > 1$. For $0 < \alpha \leq 1$ it is neither necessary nor sufficient (see Exercises 5.2.3 and 5.2.6). We shall return to this point after proving the theorem.

Proof of Theorem 5.2.5 (i) We use the reproducing kernels k_w as test functions. By (5.2), applied with $f := k_w$, there exists a constant C such that

$$\int_{\mathbb{D}} |k_w|^2\,d\mu(z) \leq C\|k_w\|_{\mathcal{D}}^2 \qquad (w \in \mathbb{D}). \tag{5.9}$$

We shall prove that this implies (5.7).

We may as well limit our attention to those arcs I with $|I| \leq 1/4$. Fix such an arc, and let w be the point in $\mathbb{D}$ such that $|w| = 1 - |I|$ and $w/|w|$ is the midpoint of I. Elementary estimates show that, if $z \in S(I)$, then $|1 - \overline{w}z| \leq 3|I|$. Hence

$$|k_w(z)| = \Big|\frac{1}{\overline{w}z}\log\Big(\frac{1}{1-\overline{w}z}\Big)\Big| \geq \log\Big|\frac{1}{1-\overline{w}z}\Big| \geq \log\frac{1}{3|I|} \qquad (z \in S(I)),$$

and consequently

$$\int_{\mathbb{D}} |k_w|^2\,d\mu \geq \int_{S(I)} |k_w|^2\,d\mu \geq \mu(S(I))\Big(\log\frac{1}{3|I|}\Big)^2.$$

Also, using the reproducing property of k_w and the fact that $|w| = 1 - |I| \geq 3/4$, we have

$$\|k_w\|_{\mathcal{D}}^2 = k_w(w) = \frac{1}{|w|^2}\log\Big(\frac{1}{1-|w|^2}\Big) \leq \frac{16}{9}\log\frac{1}{|I|}.$$

Feeding these estimates into (5.9), we obtain

$$\mu(S(I))\Big(\log\frac{1}{3|I|}\Big)^2 \leq \frac{16}{9}C\log\frac{1}{|I|}.$$

This implies (5.7).

(ii) It is enough to show that (5.8) implies (5.4). In establishing (5.4), we can restrict our attention to those w with $1/2 < |w| < 1$, since the supremum over the remaining w is clearly finite. For convenience, we extend the domain of definition of ϕ to $\mathbb{R}^+$ by setting $\phi(t) := \phi(2\pi)$ for $t > 2\pi$.

Fix w with $1/2 < |w| < 1$. Then

$$\begin{aligned}\int_{\mathbb{D}} \log \frac{2}{|1-\overline{w}z|}\, d\mu(z) &= \int_{z\in\mathbb{D}} \int_{t=|1-\overline{w}z|}^{2} \frac{1}{t}\, dt\, d\mu(z)\\ &= \int_{t=0}^{2} \int_{\{z\in\mathbb{D}:|1-\overline{w}z|\le t\}} \frac{1}{t}\, d\mu(z)\, dt\\ &= \int_{t=0}^{2} \mu\Big(\{z\in\mathbb{D} : |1-\overline{w}z| \le t\}\Big)\frac{1}{t}\, dt.\end{aligned}$$

Now $\{z \in \mathbb{D} : |1-\overline{w}z| \le t\} = \{z \in \mathbb{D} : |z - 1/\overline{w}| \le t/|w|\}$. This set is contained in $S(I)$ for some arc I with $|I| = 8t$, so by (5.8) we have

$$\mu\Big(\{z\in\mathbb{D} : |1-\overline{w}z| \le t\}\Big) \le C\phi(8t),$$

where C is a constant independent of w and t. Feeding this estimate into the previous equality, we obtain

$$\int_{\mathbb{D}} \log \frac{2}{|1-\overline{w}z|}\, d\mu(z) \le \int_{t=0}^{2} C\phi(8t)\frac{1}{t}\, dt = \int_{s=0}^{16} C\frac{\phi(s)}{s}\, ds < \infty.$$

This gives (5.4). □

As mentioned just after the statement of Theorem 5.2.5, there is a small gap between the necessary condition and the sufficient condition given in that theorem. Our final result in this section provides a condition that is simultaneously necessary and sufficient for μ to be a Carleson measure for $\mathcal{D}$, albeit at the cost of some additional complication. As usual, we write c for logarithmic capacity.

Theorem 5.2.6 *Let μ be a finite positive Borel measure on $\mathbb{D}$. Then μ is a Carleson measure for $\mathcal{D}$ if and only if there exists a constant A such that, for every finite set of disjoint closed subarcs $I_1, \dots, I_n$ of $\mathbb{T}$,*

$$\mu\Big(\cup_{k=1}^n S(I_k)\Big) \le Ac\Big(\cup_{k=1}^n I_k\Big). \tag{5.10}$$

In this theorem, the condition (5.10) cannot be weakened to hold merely for single arcs. Indeed, since $c(I) \asymp 1/\log(1/|I|)$ for all small arcs I (see §2.4), the weakened condition would be equivalent to (5.7), which is not sufficient to ensure that μ be a Carleson measure for $\mathcal{D}$ (see Exercise 5.2.6).

Proof of Theorem 5.2.6 We first prove necessity. Let μ be a Carleson measure for $\mathcal{D}$, so there exists a constant C such that

$$\int_{\mathbb{D}} |f|^2 \, d\mu \leq C\|f\|_{\mathcal{D}}^2 \qquad (f \in \mathcal{D}). \tag{5.11}$$

Let $I_1, \ldots, I_n$ be disjoint closed subarcs of $\mathbb{T}$. Set $F := \cup_1^n I_j$, and let ν be the equilibrium measure for F. Define $f_\nu : \mathbb{D} \to \mathbb{C}$ by

$$f_\nu(z) := \int_F \log\Big(\frac{2}{1 - \overline{\zeta}z}\Big) \, d\nu(\zeta) \qquad (z \in \mathbb{D}).$$

By Lemma 3.4.2 we have $f_\nu \in \mathcal{D}$ with $\mathcal{D}(f_\nu) = 1/c(F) - \log 2$. It follows that

$$\|f_\nu\|_{\mathcal{D}}^2 = \|f_\nu\|_{H^2}^2 + \mathcal{D}(f_\nu) \leq |f_\nu(0)|^2 + 2\mathcal{D}(f_\nu) \leq 2/c(F).$$

Also, by Lemma 3.4.2, we have $\operatorname{Re} f_\nu \geq 0$ on $\mathbb{D}$, and further $\operatorname{Re} f_\nu^* \geq 1/c(F)$ q.e. on F, hence a.e. on F with respect to Lebesgue measure. Therefore, if $z \in S(I_j)$, then

$$\begin{aligned}
\operatorname{Re} f_\nu(z) &\geq \frac{1}{2\pi} \int_{\mathbb{T}} \frac{1 - |z|^2}{|\zeta - z|^2} \operatorname{Re} f_\nu^*(\zeta) \, |d\zeta| \\
&\geq \frac{1}{2\pi} \int_{I_j} \frac{1 - |z|^2}{|\zeta - z|^2} \frac{1}{c(F)} \, |d\zeta| \\
&\geq \frac{1}{2\pi} \int_{I_j} \frac{|I_j|}{(2|I_j|)^2} \frac{1}{c(F)} \, |d\zeta| \\
&= \frac{1}{8\pi c(F)}.
\end{aligned}$$

Hence

$$\int_{\mathbb{D}} |f_\nu|^2 \, d\mu \geq \int_{\cup_1^n S(I_j)} |f_\nu|^2 \, d\mu \geq \frac{\mu\big(\cup_1^n S(I_j)\big)}{64\pi^2 c(F)^2}.$$

Combining all these estimates, we deduce that

$$\frac{\mu\big(\cup_1^n S(I_j)\big)}{64\pi^2 c(F)^2} \leq C \frac{2}{c(F)},$$

or in other words, that

$$\mu\big(\cup_1^n S(I_j)\big) \leq (128\pi^2 C) c\big(\cup_1^n I_j\big).$$

Therefore (5.10) holds, with $A = 128\pi^2 C$.

Now we prove sufficiency. Let μ be a finite positive measure on $\mathbb{D}$ satisfying the condition (5.10). This condition implies (see Exercise 5.2.2) that, writing $S(U)$ for the union of the boxes $S(I)$ for all subarcs I of U, we have

$$\mu(S(U)) \leq Ac(U) \qquad (\text{open } U \subset \mathbb{T}). \tag{5.12}$$

Let $f \in \mathcal{D}$. By Theorem 3.1.3, we may write $f = Cg$, where $g \in L^2(\mathbb{A})$ with $\|g\|_{L^2(\mathbb{A})} \leq \|f\|_{\mathcal{D}}$. Then

$$\begin{aligned}\int_{\mathbb{D}} |f(z)|^2 \, d\mu(z) &= \int_{\mathbb{D}} |Cg(z)|^2 \, d\mu(z) \\ &= \int_{\mathbb{D}} \int_{t=0}^{|Cg(z)|} dt^2 \, d\mu(z) \\ &= \int_{t=0}^{\infty} \int_{\{z:|Cg(z)|>t\}} d\mu(z) \, dt^2 \\ &= \int_0^{\infty} \mu\{|Cg| > t\} \, dt^2.\end{aligned}$$

We therefore need to estimate $\mu\{|Cg| > t\}$. For this, we use Theorem 3.2.4. In the notation of that theorem, we have

$$|Cg(z)| \leq \Big(1 + \frac{|z - \zeta|}{1 - |z|}\Big)\widetilde{C}g(\zeta) \qquad (z \in \mathbb{D}, \ \zeta \in \mathbb{T}).$$

Hence, if $|Cg(z)| > t$, then $\widetilde{C}g(\zeta) > t/3$ for all ζ in the arc with center $z/|z|$ and length $2(1 - |z|)$. Thus $\{|Cg| > t\} \subset S(\{\widetilde{C}g > t/3\})$. By (5.12), it follows that

$$\mu\{|Cg| > t\} \leq \mu(S(\{\widetilde{C}g > t/3\})) \leq Ac(\widetilde{C}g > t/3).$$

Feeding this information back into our previous estimate, we obtain

$$\int_{\mathbb{D}} |f|^2 \, d\mu \leq \int_0^{\infty} Ac(\widetilde{C}g > t/3) \, dt^2 = 9A \int_0^{\infty} c(\widetilde{C}g > s) \, ds^2.$$

By the strong-type inequality for capacity, Theorem 3.3.5,

$$\int_0^{\infty} c(\widetilde{C}g > s) \, ds^2 \leq B\|g\|^2_{L^2(\mathbb{A})},$$

where B is an absolute constant. Hence, finally,

$$\int_{\mathbb{D}} |f|^2 \, d\mu \leq 9AB\|g\|^2_{L^2(\mathbb{A})} \leq 9AB\|f\|^2_{\mathcal{D}}.$$

Thus μ is a Carleson measure for $\mathcal{D}$, satisfying (5.11) with $C = 9AB$. □

Exercises 5.2

1. Prove the following inequality, used in the proof of Theorem 5.2.3:

$$|k(z, w)| \leq 2 \log \frac{2}{|1 - \overline{w}z|} + \pi \qquad (z, w \in \mathbb{D}).$$

2. Show that the conditions (5.10) and (5.12) are equivalent to one another.

3. Let μ be a finite positive measure on $[0, 1)$. Using Theorem 5.2.6, show that μ is a Carleson measure for $\mathcal{D}$ if and only if
$$\mu\big((1-t, 1)\big) = O\Big(\frac{1}{\log(1/t)}\Big) \qquad (t \to 0).$$
Deduce that the necessary condition (5.7) in Theorem 5.2.5 (i) is sharp.
4. Give an example of a Carleson measure μ for $\mathcal{D}$ that does not satisfy (5.4). [Hint: Use Exercise 3.]
5. Let μ be a finite positive measure on $\mathbb{D}$.
   (i) Let $f \in \mathcal{D}$, say $f(z) := \sum_k a_k z^k$, and set $\widetilde{f}(z) := \sum_k |a_k| z^k$. Show that $\int_{\mathbb{D}} |f|^2\, d\mu \leq \int_{\mathbb{D}} |\widetilde{f}|^2\, d\widetilde{\mu}$, where $\widetilde{\mu}$ is the measure on $[0, 1)$ given by
   $$\widetilde{\mu}(S) := \mu(\{z : |z| \in S\}) \qquad (S \subset [0, 1)).$$
   (ii) Deduce that, if
   $$\mu\big(\{z : |z| > 1 - t\}\big) = O\Big(\frac{1}{\log(1/t)}\Big) \qquad (t \to 0),$$
   then μ is a Carleson measure for $\mathcal{D}$. [Hint: Apply Exercise 3 to $\widetilde{\mu}$.]
6. Let $\phi : (0, 2\pi] \to \mathbb{R}^+$ be a continuous strictly increasing function such that $\phi(x)/x$ is strictly decreasing and $\int_0^{2\pi} (\phi(x)/x)\, dx = \infty$. The purpose of this question is to construct a finite positive measure μ on $\mathbb{D}$ that satisfies (5.8) but is not a Carleson measure for $\mathcal{D}$, thereby showing that Theorem 5.2.5 (ii) is sharp.
   (i) For each $n \geq 0$, set $l_n := \phi^{-1}(2^{-n})$. Show that $l_{n+1} < l_n/2$ for all n and that $\sum_n 2^{-n} \log(1/l_n) = \infty$.
   (ii) Let E be the Cantor subset of $\mathbb{T}$ associated to the sequence (l_n) (thus $E = \cap_n E_n$, where E_n is a union of 2^n disjoint arcs each of length l_n). Show that E is of logarithmic capacity zero. [Use Theorem 2.3.5.]
   (iii) Let σ be the Cantor–Lebesgue measure associated to E, namely the unique probability measure supported on E giving weight 2^{-n} to each of the arcs making up E_n. Show that $\sigma(I) \leq 4\phi(|I|)$ for each arc I.
   (iv) Let (δ_n) be any decreasing sequence in $(0, 1)$. Let μ_n be the measure on $\mathbb{D}$ defined by $\mu_n((1-\delta_n)B) := \sigma(B)$ (so μ_n is just σ, scaled to live on the slightly smaller circle $|z| = 1 - \delta_n$). Finally, let $\mu := \sum_{n\geq 0} 2^{-n}\mu_n$. Show that, for each arc I, we have $\mu(S(I)) \leq 2\sigma(I)$, and deduce that μ satisfies (5.8).
   (v) Show that the set $E_{\delta_n} := \{\zeta \in \mathbb{T} : d(\zeta, E) \leq \delta_n\}$ is a union of closed arcs $I_1, \ldots, I_k$ satisfying $\mu(\cup_1^k S(I_j)) \geq 2^{-n}$. Deduce that, if (5.10) holds, then necessarily $2^{-n} \leq Ac(E_{\delta_n})$. Conclude that, if (δ_n) is chosen to

decrease fast enough, then (5.10) fails and μ is not a Carleson measure for $\mathcal{D}$.

5.3 Pick interpolation

The classic Pick interpolation problem is to determine whether, given distinct points $z_1, \dots, z_n \in \mathbb{D}$ and points $w_1, \dots, w_n \in \overline{\mathbb{D}}$, there exists a holomorphic map $h : \mathbb{D} \to \overline{\mathbb{D}}$ such that $h(z_j) = w_j$ for all j. As is well known, a necessary and sufficient condition for h to exist is that the $n \times n$ matrix (a_{ij}) defined by

$$a_{ij} := \frac{1 - w_i \overline{w}_j}{1 - z_i \overline{z}_j} \qquad (i, j \in \{1, \dots, n\}) \tag{5.13}$$

be positive semi-definite.

In this section, we shall prove an analogue of this result for the Dirichlet space. To formulate this analogue, we first re-write the condition that h maps $\mathbb{D}$ into $\overline{\mathbb{D}}$ simply as $\|h\|_{H^\infty} \leq 1$. Now, as was mentioned in the introduction to the chapter, the natural counterpart of H^∞ in the Dirichlet-space setting is the multiplier algebra $\mathcal{M}(\mathcal{D})$. If we adopt this point of view, then the Dirichlet-space analogue of the Pick problem can be stated as follows. *Given distinct points $z_1, \dots, z_n \in \mathbb{D}$ and points $w_1, \dots, w_n \in \overline{\mathbb{D}}$, determine whether there exists $h \in \mathcal{M}(\mathcal{D})$ with $\|h\|_{\mathcal{M}(\mathcal{D})} \leq 1$ such that $h(z_j) = w_j$ for all j.* This is the problem that we are going to solve.

Theorem 5.3.1 *Let $z_1, \dots, z_n \in \mathbb{D}$ be distinct points and let $w_1, \dots, w_n \in \overline{\mathbb{D}}$. Then there exists $h \in \mathcal{M}(\mathcal{D})$ with $\|h\|_{\mathcal{M}(\mathcal{D})} \leq 1$ such that $h(z_j) = w_j$ for all j if and only if the $n \times n$ matrix (a_{ij}) defined by*

$$a_{ij} := (1 - \overline{w}_i w_j)\langle k_{z_i}, k_{z_j} \rangle_{\mathcal{D}} \qquad (i, j \in \{1, \dots, n\}) \tag{5.14}$$

is positive semi-definite.

The condition (5.14) is the natural analogue of (5.13). Indeed, if we replace $\langle k_{z_i}, k_{z_j} \rangle_{\mathcal{D}}$ by the inner product in H^2 of the H^2-reproducing kernels for z_i, z_j, then we obtain exactly (5.13).

Proof of Theorem 5.3.1 The 'only if' is very simple. Assume that h exists. Let $M_h : \mathcal{D} \to \mathcal{D}$ be the multiplication operator $M_h(f) := hf$, and let M_h^* be its Hilbert-space adjoint. Then we have $\|M_h^*\| = \|M_h\| = \|h\|_{\mathcal{M}(\mathcal{D})} \leq 1$. Also, by Theorem 5.1.4, each reproducing kernel k_{z_j} is an eigenvector of M_h^* with eigenvalue $\overline{h(z_j)} = \overline{w}_j$. Hence, for all $\lambda_1, \ldots, \lambda_n \in \mathbb{C}$, we have

$$\begin{aligned}
0 &\leq \Big\|\sum_{j=1}^n \lambda_j k_{z_j}\Big\|_{\mathcal{D}}^2 - \Big\|M_h^*\big(\sum_{j=1}^n \lambda_j k_{z_j}\big)\Big\|_{\mathcal{D}}^2 \\
&= \Big\|\sum_{j=1}^n \lambda_j k_{z_j}\Big\|_{\mathcal{D}}^2 - \Big\|\sum_{j=1}^n \lambda_j \overline{w}_j k_{z_j}\Big\|_{\mathcal{D}}^2 \\
&= \sum_{i=1}^n \sum_{j=1}^n \lambda_i \overline{\lambda}_j (1 - \overline{w}_i w_j)\langle k_{z_i}, k_{z_j}\rangle_{\mathcal{D}}.
\end{aligned}$$

This shows that the matrix with entries $a_{ij} := (1 - \overline{w}_i w_j)\langle k_{z_i}, k_{z_j}\rangle_{\mathcal{D}}$ is positive semi-definite.

The proof of the 'if' is a bit more involved. We shall need the following auxiliary result from operator theory.

Lemma 5.3.2 (Parrott's lemma) *Let $\mathcal{H}$ be a finite-dimensional Hilbert space. Let R be a rank-one operator on $\mathcal{H}$ and let A be an arbitrary operator on $\mathcal{H}$. Then*

$$\min_{\lambda \in \mathbb{C}} \|A + \lambda R\| = \max\{\|AP\|, \|QA\|\}, \tag{5.15}$$

where P and Q are the orthogonal projections of $\mathcal{H}$ onto $\ker R$ *and* $(\operatorname{im} R)^\perp$ *respectively.*

Proof A simple compactness argument shows that the minimum is attained. Suppose that it is attained at $\lambda = \lambda_0$ say, and set $B := A + \lambda_0 R$. Clearly

$$\min_{\lambda \in \mathbb{C}} \|A + \lambda R\| = \min_{\lambda \in \mathbb{C}} \|B + \lambda R\| = \|B\|.$$

Also, since $RP = QR = 0$, we have $AP = BP$ and $QA = QB$. Thus, with the assumption that $\|B + \lambda R\| \geq \|B\|$ for all $\lambda \in \mathbb{C}$, we need to prove that

$$\|B\| = \max\{\|BP\|, \|QB\|\}.$$

As $\|P\| = \|Q\| = 1$, we certainly have $\|B\| \geq \max\{\|BP\|, \|QB\|\}$. Suppose, if possible, that this inequality is strict. We shall show that this leads to a contradiction.

Let $y_0 \in \mathcal{H}$ be a unit vector for which $\|By_0\| = \|B\|$. We claim that B attains its norm precisely on the one-dimensional subspace spanned by y_0. Indeed, the set where B attains its norm is the subspace $\ker(B^*B - \|B\|^2 I)$, and if this subspace had dimension two or more, then it would have a non-zero intersection with $\ker R$, on which B never attains its norm since $\|BP\| < \|B\|$. The claim is established.

We next show that

$$\langle By_0, Ry_0 \rangle = 0. \tag{5.16}$$

Fix $\theta \in \mathbb{R}$. For each $\epsilon > 0$, let $y_\epsilon \in \mathcal{H}$ be a unit vector on which $B + \epsilon e^{i\theta} R$ attains its norm. Since $\|B + \epsilon e^{i\theta} R\| \geq \|B\|$, it follows that

$$\|(B + \epsilon e^{i\theta} R)y_\epsilon\|^2 \geq \|B\|^2 \qquad (\epsilon > 0).$$

Expanding this out and dividing through by ϵ, we obtain

$$2\,\mathrm{Re}(e^{-i\theta}\langle By_\epsilon, Ry_\epsilon\rangle) + O(\epsilon) \geq 0 \qquad (\epsilon \to 0^+).$$

By compactness, there exists a sequence $\epsilon_n \to 0$ such that the corresponding vectors y_{ϵ_n} converge. The limit vector is a unit vector in $\mathcal{H}$ on which B attains its norm so, by the claim above, it is necessarily a unimodular multiple of y_0. Letting $\epsilon \to 0$ through this sequence (ϵ_n), we therefore obtain

$$2\,\mathrm{Re}(e^{-i\theta}\langle By_0, Ry_0\rangle) \geq 0.$$

As θ is arbitrary, it follows that (5.16) holds.

There are now two possibilities: either $Ry_0 = 0$ or $Ry_0 \neq 0$. In the first case, we have $y_0 \in \ker R$, which implies

$$\|By_0\| = \|BPy_0\| \leq \|BP\| < \|B\|.$$

In the second case, thanks to (5.16) we have $By_0 \in (\mathrm{im}\, R)^\perp$, which implies

$$\|By_0\| = \|QBy_0\| \leq \|QB\| < \|B\|.$$

Either way, B does not attain its norm on y_0, contradicting our choice of y_0. The lemma is proved. □

We now return to the proof of the 'if' part of Theorem 5.3.1. Recall that we are given $z_1, \ldots, z_n \in \mathbb{D}$ and $w_1, \ldots, w_n \in \overline{\mathbb{D}}$ such that the matrix (a_{ij}) defined in (5.14) is positive semi-definite, and our objective is to construct a multiplier h with $\|h\|_{\mathcal{M}(\mathcal{D})} \leq 1$ such that $h(z_j) = w_j$ for all j. To this end, we define an operator T on the subspace $\mathcal{H}_n$ of $\mathcal{D}$ spanned by $\{k_{z_1}, \ldots, k_{z_n}\}$ by setting

$$T(k_{z_j}) := \overline{w}_j k_{z_j} \qquad (j = 1, \ldots, n).$$

The calculation in the first part of the proof shows that the assumption that (a_{ij}) is positive semi-definite is exactly equivalent to the fact that $\|T\| \leq 1$.

Now let z_{n+1} be any point of $\mathbb{D}$ distinct from $z_1, \ldots, z_n$. Given $w \in \overline{\mathbb{D}}$, we define an operator T_w on $\mathcal{H}_{n+1} := \mathrm{span}\{k_{z_1}, \ldots, k_{z_{n+1}}\}$ by setting

$$T_w(k_{z_j}) := \begin{cases} \overline{w}_j k_{z_j}, & j = 1, \ldots, n, \\ \overline{w} k_{z_{n+1}}, & j = n+1. \end{cases}$$

Evidently T_w is an extension of T. Our aim is to show that, with an appropriate choice of w, we still have $\|T_w\| \le 1$. For this, we are going to use Lemma 5.3.2.

Fix a vector $u \in \mathcal{H}_{n+1} \ominus \mathcal{H}_n$ such that $\langle k_{z_{n+1}}, u\rangle_{\mathbb{D}} = 1$, and define a rank-one operator $R : \mathcal{H}_{n+1} \to \mathcal{H}_{n+1}$ by

$$R(f) := \langle f, u\rangle_{\mathbb{D}}\, k_{z_{n+1}} \qquad (f \in \mathcal{H}_{n+1}).$$

Notice that $T_w = T_0 + wR$ for all $w \in \mathbb{C}$. Therefore, by Lemma 5.3.2,

$$\min_{w\in\mathbb{C}} \|T_w\| = \max\{\|T_0P\|, \|QT_0\|\},$$

where P, Q are the orthogonal projections of $\mathcal{H}_{n+1}$ onto $\ker R$ and $(\operatorname{im} R)^\perp$ respectively. We shall estimate each of $\|T_0P\|$ and $\|QT_0\|$ in turn.

First we consider T_0P. This is the same thing as the restriction of T_0 to $\mathcal{H}_n$, which is exactly T. As remarked earlier, $\|T\| \le 1$, so we have $\|T_0P\| \le 1$.

Now we turn to QT_0. This is a little more complicated. Let us note first that $QT_0(k_{z_{n+1}}) = 0$. A basis for $\mathcal{H}_{n+1} \ominus \mathbb{C}k_{z_{n+1}}$ is given by

$$\widetilde{k_{z_j}} := Q(k_{z_j}) = k_{z_j} - \frac{\langle k_{z_j}, k_{z_{n+1}}\rangle_{\mathbb{D}}}{\langle k_{z_{n+1}}, k_{z_{n+1}}\rangle_{\mathbb{D}}} k_{z_{n+1}} \qquad (j = 1, \dots, n).$$

A simple calculation shows that

$$QT_0(\widetilde{k_{z_j}}) = \overline{w_j}\widetilde{k_{z_j}} \qquad (j = 1, \dots, n).$$

Thus $\|QT_0\| \le 1$ if and only if the $n \times n$ matrix (c_{ij}) is positive semi-definite, where

$$c_{ij} := (1 - \overline{w_i}w_j)\langle \widetilde{k_{z_i}}, \widetilde{k_{z_j}}\rangle_{\mathbb{D}} \qquad (i, j \in \{1, \dots, n\}).$$

Now a further calculation shows that

$$\begin{aligned}
\langle \widetilde{k_{z_i}}, \widetilde{k_{z_j}}\rangle_{\mathbb{D}} &= \langle Q(k_{z_i}), Q(k_{z_j})\rangle_{\mathbb{D}} = \langle k_{z_i}, Q(k_{z_j})\rangle_{\mathbb{D}} \\
&= \langle k_{z_i}, k_{z_j}\rangle_{\mathbb{D}} - \frac{\overline{\langle k_{z_j}, k_{z_{n+1}}\rangle_{\mathbb{D}}}\langle k_{z_i}, k_{z_{n+1}}\rangle_{\mathbb{D}}}{\langle k_{z_{n+1}}, k_{z_{n+1}}\rangle_{\mathbb{D}}} \\
&= \langle k_{z_i}, k_{z_j}\rangle_{\mathbb{D}}\Big(1 - \frac{\langle k_{z_i}, k_{z_{n+1}}\rangle_{\mathbb{D}}\langle k_{z_{n+1}}, k_{z_j}\rangle_{\mathbb{D}}}{\langle k_{z_i}, k_{z_j}\rangle_{\mathbb{D}}\langle k_{z_{n+1}}, k_{z_{n+1}}\rangle_{\mathbb{D}}}\Big).
\end{aligned}$$

Consequently, $c_{ij} = a_{ij}b_{ij}$, where

$$a_{ij} := (1 - \overline{w_i}w_j)\langle k_{z_i}, k_{z_j}\rangle_{\mathbb{D}}$$

and

$$b_{ij} := 1 - \frac{\langle k_{z_i}, k_{z_{n+1}}\rangle_{\mathbb{D}}\langle k_{z_{n+1}}, k_{z_j}\rangle_{\mathbb{D}}}{\langle k_{z_i}, k_{z_j}\rangle_{\mathbb{D}}\langle k_{z_{n+1}}, k_{z_{n+1}}\rangle_{\mathbb{D}}}.$$

The matrix (a_{ij}) is positive semi-definite, by assumption, and the matrix (b_{ij})

was shown to be positive semi-definite in Lemma 4.2.6. Their Hadamard product (c_{ij}) is therefore positive semi-definite by the Schur product theorem (see Theorem C.2.2 in Appendix C), and we conclude that $\|QT_0\| \leq 1$, as we wanted.

In summary, we have shown that there exists $w_{n+1} \in \mathbb{C}$ such that the extended operator $T_{w_{n+1}}$ is still a contraction. Moreover, as $\overline{w}_{n+1}$ is an eigenvalue of a contraction, necessarily $w_{n+1} \in \overline{\mathbb{D}}$.

We now iterate this argument. Let (z_j) be a dense sequence of distinct points in $\mathbb{D}$, beginning with the given n-tuple $z_1, \dots, z_n$. By applying the above procedure repeatedly, we can construct successively $w_{n+1}, w_{n+2}, \dots \in \overline{\mathbb{D}}$ such that, for each $m \geq n$, the operator T defined on the span of $\{k_{z_j} : 1 \leq j \leq m\}$ by

$$T(k_{z_j}) := \overline{w}_j k_j \qquad (1 \leq j \leq m)$$

is always a contraction.

As the set $\{k_{z_j} : j \geq 1\}$ spans a dense subspace of $\mathcal{D}$, the operator T extends by continuity to the whole of $\mathcal{D}$, still remaining a contraction. By construction, all the reproducing kernels k_{z_j} are eigenvectors of T. As (z_j) is dense in $\mathbb{D}$ and the map $z \mapsto k_z : \mathbb{D} \to \mathcal{D}$ is continuous (see Exercise 5.3.1), it follows that k_z is an eigenvector of T for all $z \in \mathbb{D}$. By Theorem 5.1.4, there exists a multiplier h of $\mathcal{D}$ such that $T = M_h^*$. Clearly this h satisfies $\|h\|_{\mathcal{M}(\mathcal{D})} \leq 1$ and $h(z_j) = w_j$ $(j = 1, \dots, n)$. □

The 'if' part of the theorem is sometimes expressed by saying that the Dirichlet space has the *Pick property*. Strictly speaking, this is a property not just of the space, but of the particular norm $\|\cdot\|_{\mathcal{D}}$, or equivalently, of the reproducing kernel k_w. Indeed, if we were to replace $\|\cdot\|_{\mathcal{D}}$ by the equivalent Hilbert-space norm on $\mathcal{D}$ given by

$$\|f\|^2 := |f(0)|^2 + \mathcal{D}(f) \qquad (f \in \mathcal{D}), \tag{5.17}$$

then $(\mathcal{D}, \|\cdot\|)$ would no longer have the Pick property (see Exercise 5.3.3).

Exercises 5.3

1. As usual, let k_w be the reproducing kernel for $\mathcal{D}$ at w. Prove that the map $w \mapsto k_w : \mathbb{D} \to \mathcal{D}$ is continuous.
2. Show that, if $h \in \mathcal{M}(\mathcal{D})$ with $\|h\|_{\mathcal{M}(\mathcal{D})} \leq 1$, then

$$\left|\frac{h(z_1) - h(z_2)}{1 - \overline{h(z_1)}h(z_2)}\right|^2 \leq 1 - \frac{|\langle k_{z_1}, k_{z_2}\rangle_{\mathcal{D}}|^2}{\|k_{z_1}\|_{\mathcal{D}}^2 \|k_{z_2}\|_{\mathcal{D}}^2} \qquad (z_1, z_2 \in \mathbb{D}).$$

3. Let $\|\cdot\|$ be the Hilbert-space norm on $\mathcal{D}$ given by (5.17).

(i) Show that the associated reproducing kernel k_w is given by

$$k_w(z) = 1 + \log\Big(\frac{1}{1-\overline{w}z}\Big) \qquad (w \in \mathbb{D},\ z \in \mathbb{D}).$$

(ii) Show that the pairs $(z_1, z_2) := (0, z_0)$ and $(w_1, w_2) := (0, w_0)$ satisfy the condition (5.14) with respect to this kernel if and only if

$$(1 - |w_0|^2)\Big(1 + \log\Big(\frac{1}{1-|z_0|^2}\Big)\Big) \geq 1. \tag{5.18}$$

(iii) Let $h(z) := \sum_{k\geq 0} a_k z^k$ be a multiplier of $\mathcal{D}$ whose multiplier norm with respect to $\|\cdot\|$ is at most 1. Show that $\sum_{k\geq 0}(k+1)|a_k|^2 \leq 1$. Deduce that, if $h(0) = 0$ and $h(z_0) = w_0$, then

$$|w_0| \leq \frac{1}{\sqrt{2}}\frac{|z_0|}{1-|z_0|} \qquad (z \in \mathbb{D}). \tag{5.19}$$

(iv) Show that there exists a pair (z_0, w_0) satisfying (5.18) but not (5.19). Conclude that $(\mathcal{D}, \|\cdot\|)$ does not have the Pick property.

5.4 Zeros of multipliers

If $f \in H^2$ and $f \not\equiv 0$, then its zeros satisfy the Blaschke condition, and the corresponding Blaschke product is a function in H^∞ with the same zeros as f. The following theorem is the analogous result for the Dirichlet space.

Theorem 5.4.1 *Given $f \in \mathcal{D}$, there exists $h \in \mathcal{M}(\mathcal{D})$ with the same zero set.*

We note straightaway the following consequence.

Corollary 5.4.2 *The union of two zero sets for $\mathcal{D}$ is again a zero set.*

Proof By the theorem, the two sets in question are the zero sets of multipliers h_1, h_2. Their union is the zero set of the product h_1h_2, which belongs to $\mathcal{D}$. □

We now turn to the proof of Theorem 5.4.1. In contrast to the case of Hardy spaces, we cannot simply build h using a Blaschke product: indeed the Dirichlet space contains no infinite Blaschke products (see Exercise 4.1.1). Instead, we use the Pick interpolation theorem from the previous section.

Proof of Theorem 5.4.1 We can suppose that $\|f\|_{\mathcal{D}} = 1$. Let $(z_j)_{j\geq 1}$ be the zeros of f, counted according to multiplicity. Fix $z_0 \in \mathbb{D}$ such that $f(z_0) \neq 0$.

Set $w_0 := f(z_0)/\|k_{z_0}\|_{\mathcal{D}}^2$ and $w_j := 0$ $(j \geq 1)$. Then, for each $n \geq 1$ and each $(n+1)$-tuple $\lambda_0, \lambda_1, \dots, \lambda_n \in \mathbb{C}$, we have

$$\begin{aligned}\sum_{i=0}^{n}\sum_{j=0}^{n} \lambda_i \overline{\lambda}_j (1 - \overline{w}_i w_j)\langle k_{z_i}, k_{z_j}\rangle_{\mathcal{D}} &= \Big\|\sum_{j=0}^{n} \lambda_j k_{z_j}\Big\|^2 - |\lambda_0|^2 |f(z_0)|^2 \\ &= \Big\|\sum_{j=0}^{n} \lambda_j k_{z_j}\Big\|^2 - \Big|\langle f, \sum_{j=0}^{n} \lambda_j k_{z_j}\rangle_{\mathcal{D}}\Big|^2 \geq 0.\end{aligned}$$

By Theorem 5.3.1, there exists $h_n \in \mathcal{M}(\mathcal{D})$ with $\|h_n\|_{\mathcal{M}(\mathcal{D})} \leq 1$ such that $h_n(z_0) = f(z_0)/\|k_{z_0}\|_{\mathcal{D}}$ and $h_n(z_j) = 0$ $(j = 1, \dots, n)$. Then, for all n, we have $\|h_n\|_{H^\infty} \leq \|h_n\|_{\mathcal{M}(\mathcal{D})} \leq 1$, so $(h_n)_{n \geq 1}$ is a normal family, and some subsequence converges locally uniformly on $\mathbb{D}$. The limit function h is also a multiplier with $\|h\|_{\mathcal{M}(\mathcal{D})} \leq 1$ (see Exercise 5.1.2), and clearly $h(z_0) = f(z_0)/\|k_{z_0}\|_{\mathcal{D}} \neq 0$ and $h(z_j) = 0$ $(j \geq 1)$.

This is not quite the end of the story, since it is possible that h has more zeros than f. However, we can simply divide them out to obtain the desired function (see Exercise 5.4.1). □

We shall give another proof of this theorem, in a slightly more general form, at the end of Chapter 8 (see Corollary 8.3.10). Also in Chapter 9, we shall establish an analogue for boundary zero sets (see Theorem 9.2.6).

Exercises 5.4

1. Let $h \in \mathrm{Hol}(\mathbb{D})$ and B be a Blaschke product. Show that, if $hB \in \mathcal{M}(\mathcal{D})$, then also $h \in \mathcal{M}(\mathcal{D})$ and $\|h\|_{\mathcal{M}(\mathcal{D})} \leq \|hB\|_{\mathcal{M}(\mathcal{D})}$. [Use Theorem 4.1.3.]
2. Let Z be a zero set for $\mathcal{D}$ and let $z_0 \in \mathbb{D} \setminus Z$. Define

$$\begin{aligned}c_{\mathcal{D}} &:= \inf\{\|f\|_{\mathcal{D}} : f \in \mathcal{D},\ f|_Z = 0,\ f(z_0) = 1\}, \\ c_{\mathcal{M}(\mathcal{D})} &:= \inf\{\|h\|_{\mathcal{M}(\mathcal{D})} : h \in \mathcal{M}(\mathcal{D}),\ h|_Z = 0,\ h(z_0) = 1\}.\end{aligned}$$

   Prove that

$$c_{\mathcal{M}(\mathcal{D})}/c_{\mathcal{D}} = \|k_{z_0}\|_{\mathcal{D}}.$$

   [The proof of Theorem 5.4.1 gives the inequality $\leq$. For the reverse inequality, consider $f := hk_{z_0}/\|k_{z_0}\|_{\mathcal{D}}^2$.]
3. [Another proof of Theorem 4.2.1.] Let (z_n) be a sequence in $\mathbb{D} \setminus \{0\}$ such that

$$\sum_n \Big\{\log\Big(\frac{1}{1 - |z_n|}\Big)\Big\}^{-1} < \infty.$$

(i) For each n, show that there exists $h_n \in \mathcal{M}(\mathcal{D})$ with $\|h_n\|_{\mathcal{M}(\mathcal{D})} \leq 1$ such that $h_n(z_n) = 0$ and

$$|h_n(0)| \geq 1 - \frac{|z_n|^2}{\log(\frac{1}{1-|z_n|^2})}.$$

(ii) Let $h := \prod_n h_n$. Show that $h \in \mathcal{M}(\mathcal{D})$ and $h(z_n) = 0$ for all n, but $h(0) \neq 0$.

Notes on Chapter 5

§5.1

Theorem 5.1.6 is based on an example of Taylor [119]. Another proof that not every bounded function in $\mathcal{D}$ is a multiplier can be found in [117]. Exercise 5.1.3 is a result of Brown and Shields [24].

§5.2

The notion of Carleson measure (for H^p-spaces) was introduced by Carleson in his solution of the corona problem [28]. Theorem 5.2.2 is based on a result of Arcozzi, Rochberg and Sawyer in [10]. Theorem 5.2.5 (ii) was obtained by Wynn in [124] under the additional assumption that ϕ is a concave function, and the sharpness result of Exercise 5.2.6 was also established in [124]. The proofs given here are from [42]. Theorem 5.2.6 is due to Stegenga [117]. As outlined in Exercise 5.2.3, Stegenga's theorem leads to a simple characterization of those Carleson measures for $\mathcal{D}$ supported on $[0, 1)$, but this result can also be obtained directly, see [114]. Exercise 5.2.5 is taken from [102].

There are a number of other characterizations of Carleson measures for $\mathcal{D}$, though none of them are particularly simple. For example, Arcozzi, Rochberg and Sawyer have shown that a finite positive measure μ on $\mathbb{D}$ is a Carleson measure for $\mathcal{D}$ if and only if there exists a constant C such that, for all closed arcs I in $\mathbb{T}$,

$$\int_{S(I)} \int_{S(I)} \operatorname{Re} k(z, w)\, d\mu(w)\, d\mu(z) \leq C\mu(S(I)).$$

For a discussion of this and other characterizations, we refer to the articles [9, 10, 11, 64].

§5.3

Theorem 5.3.1 is due to Agler [2], and Lemma 5.3.2 is a special case of a theorem of Parrott [89].

In fact the Dirichlet space enjoys a stronger property, the so-called complete Pick property, which corresponds to a version of Theorem 5.3.1 for matrix-valued multipliers. Shimorin [116] has shown that the weighted Dirichlet space $\mathcal{D}_w$ has the complete Pick property whenever w is a positive superharmonic function (even in the absence of an explicit formula for the reproducing kernel). The book of Agler and McCarthy [4] contains a detailed account of the complete Pick property for general reproducing kernels.

§5.4

The proof of Theorem 5.4.1 is taken from the preprint of Marshall and Sundberg [76], which is also the source for Exercises 5.4.2 and 5.4.3.

The main purpose of [76] was to establish a characterization of interpolating sequences for $\mathcal{D}$. This result, which was also proved independently by Bishop [17] and later simplified by Bøe [18], can be stated as follows. Given a sequence of distinct points $(z_n)_{n\geq 1}$ in $\mathbb{D}$, we have

$$\{(h(z_1), h(z_2), h(z_3), \dots) : h \in \mathcal{M}(\mathcal{D})\} = \ell^\infty$$

if and only if $\sum_n \|k_{z_n}\|_{\mathcal{D}}^{-2}\delta_{z_n}$ is a Carleson measure for $\mathcal{D}$ and

$$\sup_{\substack{m,n \\ m\neq n}} \frac{|\langle k_{z_n}, k_{z_m}\rangle_{\mathcal{D}}|}{\|k_{z_n}\|_{\mathcal{D}}\|k_{z_m}\|_{\mathcal{D}}} < 1.$$

For a detailed treatment, we refer to the book of Seip [109].

6

Conformal invariance

Back in §1.4, we saw that if ϕ is a Möbius automorphism of the unit disk, then $\mathcal{D}(f \circ \phi) = \mathcal{D}(f)$ for all $f \in \mathcal{D}$. Despite its simplicity, this observation has already proved useful, notably in the proof of Theorem 4.1.3. In this chapter, we start by showing that this Möbius-invariance property essentially characterizes the Dirichlet space. We then go on to study the problem of determining which other maps $\phi : \mathbb{D} \to \mathbb{D}$ leave $\mathcal{D}$ invariant in the sense that $f \in \mathcal{D} \Rightarrow f \circ \phi \in \mathcal{D}$. This leads to the notion of composition operators on $\mathcal{D}$, to which we provide a brief introduction.

6.1 Möbius invariance

Let $\mathrm{Aut}(\mathbb{D})$ denote the group of holomorphic automorphisms of the unit disk, namely the Möbius maps of the form

$$\phi(z) := e^{i\theta}\frac{a-z}{1-\bar{a}z} \qquad (a \in \mathbb{D},\ e^{i\theta} \in \mathbb{T}).$$

There are many interesting spaces X of holomorphic functions on $\mathbb{D}$ that are Möbius invariant in the sense that

$$f \in X,\ \phi \in \mathrm{Aut}(\mathbb{D}) \Rightarrow f \circ \phi \in X.$$

Some of them are even Hilbert spaces. For example, both H^2 and $\mathcal{D}$ have this property (for the case of H^2, see Exercise 6.1.2). However, rather remarkably, the stronger property that

$$\mathcal{D}(f \circ \phi) = \mathcal{D}(f) \qquad (f \in \mathcal{D},\ \phi \in \mathrm{Aut}(\mathbb{D}))$$

essentially characterizes $\mathcal{D}$ among Hilbert space of holomorphic functions. Here is a precise statement of the theorem.

Theorem 6.1.1 *Let $\mathcal{H}$ be a vector subspace of* $\mathrm{Hol}(\mathbb{D})$ *and let* $\langle\cdot,\cdot\rangle$ *be a semi-inner product on $\mathcal{H}$. Assume the following.*

(i) *If $f \in \mathcal{H}$ and $\phi \in \mathrm{Aut}(\mathbb{D})$, then $f \circ \phi \in \mathcal{H}$ and $\langle f \circ \phi, f \circ \phi\rangle = \langle f, f\rangle$.*
(ii) *$\|f\|^2 := |f(0)|^2 + \langle f, f\rangle$ defines a Hilbert-space norm $\|\cdot\|$ on $\mathcal{H}$.*
(iii) *Convergence in this norm implies pointwise convergence in $\mathbb{D}$.*
(iv) *The space $\mathcal{H}$ contains at least one non-constant function.*

Then $\mathcal{H} = \mathcal{D}$, and there is a constant c such that $\langle f, f\rangle = c\mathcal{D}(f)$ for all $f \in \mathcal{H}$.

Proof We claim that, if $f = \sum_{k\ge0} a_k z^k \in \mathcal{H}$, then $a_k z^k \in \mathcal{H}$ for each k, with $\|a_k z^k\| \le \|f\|$. To see this, fix $k \ge 0$ and consider the sequence of functions

$$g_n(z) := \frac{1}{n}\sum_{j=1}^{n} f(e^{2\pi ij/n}z)e^{-2\pi ijk/n} \qquad (n \ge 1). \tag{6.1}$$

By assumption (i), each $g_n \in \mathcal{H}$, and satisfies $\|g_n\| \le \|f\|$. By assumption (ii), $(\mathcal{H}, \|\cdot\|)$ is a Hilbert space, so there exists a subsequence (g_{n_j}) that converges weakly to some $g \in \mathcal{H}$ with $\|g\| \le \|f\|$. For each fixed $z \in \mathbb{D}$, we have

$$\lim_{n\to\infty} g_n(z) = \frac{1}{2\pi}\int_0^{2\pi} f(e^{i\theta}z)e^{-ik\theta}\,d\theta = a_k z^k. \tag{6.2}$$

By assumption (iii), evaluation at z is a continuous linear functional with respect to the norm $\|\cdot\|$, so it follows that $g(z) = a_k z^k$ for all $z \in \mathbb{D}$. This justifies the claim.

By assumption (iv), the space $\mathcal{H}$ contains a non-constant function f_0. Then $f_0 = \sum_{k\ge0} a_k z^k$ with $a_k \ne 0$ for at least one $k > 0$. By the claim, it follows that $z^k \in \mathcal{H}$ for this k. Using assumption (i), we deduce that $\phi_t^k \in \mathcal{H}$, where $\phi_t(z) := (t - z)/(1 - tz)$ $(0 < t < 1)$. The Taylor series of ϕ_t^k is

$$\phi_t(z)^k = \phi_t(0)^k + k\phi_t(0)^{k-1}\phi_t'(0)z + \cdots = t^k + kt^{k-1}(t^2 - 1)z + \cdots.$$

In particular, the Taylor coefficient of z is non-zero so, re-applying the claim, we deduce that $z \in \mathcal{H}$. By assumption (i) again, we have $\phi_t \in \mathcal{H}$. The Taylor series of ϕ_t is

$$\phi_t(z) = t + (t - 1/t)\sum_{j\ge1} t^j z^j.$$

In particular, all the Taylor coefficients are non-zero so, using the claim yet again, we have $z^k \in \mathcal{H}$ for all $k \ge 0$.

The monomials z^k are mutually orthogonal with respect to $\langle\cdot,\cdot\rangle$. This is a simple consequence of assumption (i). Indeed, given $j, k \ge 0$ with $j \ne k$, choose θ with $e^{i(j-k)\theta} \ne 1$. By assumption (i), we then have

$$\langle z^j, z^k\rangle = \langle (e^{i\theta}z)^j, (e^{i\theta}z)^k\rangle = e^{i(j-k)\theta}\langle z^j, z^k\rangle,$$

whence $\langle z^j, z^k\rangle = 0$.

Next, we compute $\langle z^k, z^k\rangle$. For this, we use the part of the claim which says that, if $f = \sum_k a_k z^k \in \mathcal{H}$, then $\|a_k z^k\| \le \|f\|$ for all k. Applying this with $f = \phi_t$, we obtain

$$\|(t - 1/t)t^k z^k\| \le \|\phi_t\| = (|\phi_t(0)|^2 + \langle \phi_t, \phi_t\rangle)^{1/2} = (t^2 + \langle z, z\rangle)^{1/2}.$$

In particular, taking $t = 1 - 1/k$, it follows that

$$\|z^k\| \le \frac{k^2 - k}{2k - 1}(1 - 1/k)^{-k}(1 + \langle z, z\rangle)^{1/2} = O(k) \qquad (k \to \infty).$$

Thus, if $0 < t < 1$, then the series $\sum_{k\ge 0} t^k z^k$ converges absolutely in $\mathcal{H}$. This permits us to expand the following inner product in powers of z:

$$\langle 1 - t\phi_t, 1 - t\phi_t\rangle = \Big\langle \frac{1 - t^2}{1 - tz}, \frac{1 - t^2}{1 - tz}\Big\rangle = (1 - t^2)^2 \sum_{k\ge 0} t^{2k}\langle z^k, z^k\rangle \quad (0 < t < 1).$$

On the other hand, by assumption (i), we also have

$$\langle 1 - t\phi_t, 1 - t\phi_t\rangle = \langle 1 - tz, 1 - tz\rangle = \langle 1, 1\rangle + t^2\langle z, z\rangle \qquad (0 < t < 1).$$

If we compare the right-hand sides of the last two expressions and equate powers of t, then we obtain

$$\begin{aligned} \langle z, z\rangle - 2\langle 1, 1\rangle &= \langle z, z\rangle, \\ \langle z^k, z^k\rangle - 2\langle z^{k-1}, z^{k-1}\rangle + \langle z^{k-2}, z^{k-2}\rangle &= 0 \qquad (k \ge 2). \end{aligned}$$

Solving this recurrence gives $\langle z^k, z^k\rangle = k\langle z, z\rangle$ for all $k \ge 0$.

To summarize, we have now proved that $\mathcal{H}$ contains all polynomials p, and that $\langle p, p\rangle = c\mathcal{D}(p)$, where $c := \langle z, z\rangle$.

We next show that $\mathcal{D}$ is a closed subspace of $\mathcal{H}$ and that $\langle f, f\rangle = c\mathcal{D}(f)$ for all $f \in \mathcal{D}$. Given $f \in \mathcal{D}$, let f_n denote the n-th partial sum of the Taylor series of f. Since the differences $f_n - f_m$ are polynomials vanishing at 0, we have $\|f_n - f_m\|^2 = c\mathcal{D}(f_n - f_m)$ for all m, n. Thus (f_n) is a Cauchy sequence with respect to $\|\cdot\|$. By assumption (ii), there exists $\widetilde{f} \in \mathcal{H}$ such that $\|f_n - \widetilde{f}\| \to 0$. By assumption (iii), this implies that $f_n \to \widetilde{f}$ pointwise in $\mathbb{D}$, so in fact $\widetilde{f} = f$. We conclude that $f \in \mathcal{H}$ and that $\langle f, f\rangle = \lim_{n\to\infty}\langle f_n, f_n\rangle = \lim_{n\to\infty} c\mathcal{D}(f_n) = c\mathcal{D}(f)$. Since $\mathcal{D}$ is complete, it is necessarily closed in $\mathcal{H}$.

The last thing to prove is that $\mathcal{D}$ is the whole of $\mathcal{H}$. For this, we consider $\mathcal{H} \ominus \mathcal{D}$, the orthogonal complement of $\mathcal{D}$ in $\mathcal{H}$. Using assumption (i), it is easy to check that, if $f \in \mathcal{H} \ominus \mathcal{D}$, then $f \circ \phi \in \mathcal{H} \ominus \mathcal{D}$ for all ϕ of the form $\phi(z) = e^{i\theta}z$. Thus, the same argument as that given at the beginning of the proof shows that, if $f = \sum_k a_k z^k \in \mathcal{H} \ominus \mathcal{D}$, then $a_k z^k \in \mathcal{H} \ominus \mathcal{D}$ for all k. But since $a_k z^k \in \mathcal{D}$, this implies that $a_k = 0$ for all k, and so $f = 0$. Thus $\mathcal{H} \ominus \mathcal{D} = \{0\}$, and $\mathcal{D}$ is the whole of $\mathcal{H}$, as desired. □

Exercises 6.1

1. Find the Taylor series expansion of the function g_u defined in (6.1). Use this series to give another proof of (6.2).
2. Let $\phi \in \mathrm{Aut}(\mathbb{D})$, say $\phi(z) = e^{i\theta}(a - z)/(1 - \overline{a}z)$, where $|a| < 1$.
   (i) Show that, if f is holomorphic in a neighborhood of $\overline{\mathbb{D}}$, then
   $$\int_{\mathbb{T}} |f(\lambda)|\,|d\lambda| = \int_{\mathbb{T}} |(f \circ \phi)(\zeta)| \frac{1 - |a|^2}{|1 - \overline{a}\zeta|^2}\,|d\zeta|.$$
   (ii) Deduce that
   $$\|f \circ \phi\|_{H^2}^2 \leq \frac{1 + |a|}{1 - |a|} \|f\|_{H^2}^2.$$
   (iii) Deduce that the same inequality holds for all $f \in H^2$.

6.2 Composition operators

Given a holomorphic map $\phi : \mathbb{D} \to \mathbb{D}$, we write $C_\phi : \mathrm{Hol}(\mathbb{D}) \to \mathrm{Hol}(\mathbb{D})$ for the composition operator defined by

$$C_\phi(f) := f \circ \phi \qquad (f \in \mathrm{Hol}(\mathbb{D})).$$

In this section, we address the problem of determining those functions ϕ for which C_ϕ maps $\mathcal{D}$ back into $\mathcal{D}$.

Let us begin with some elementary observations. If $C_\phi(\mathcal{D}) \subset \mathcal{D}$, then the linear map $C_\phi : \mathcal{D} \to \mathcal{D}$ is automatically continuous. This is a simple consequence of the closed graph theorem. We shall write $\|C_\phi\|$ for its operator norm. Also, for this to happen, ϕ itself must belong to $\mathcal{D}$, since $\phi = C_\phi(z)$. In the other direction, we saw in §1.4 that, if $\phi : \mathbb{D} \to \mathbb{D}$ is univalent, then $C_\phi(\mathcal{D}) \subset \mathcal{D}$.

To study non-univalent ϕ, we introduce the *counting function* n_ϕ. Given a holomorphic map $\phi : \mathbb{D} \to \mathbb{D}$ and $w \in \mathbb{D}$, we write $n_\phi(w)$ for the number (possibly infinite) of solutions of the equation $\phi(z) = w$, counted according to multiplicity. If ϕ is non-constant, then n_ϕ is a lower semicontinuous function on $\mathbb{D}$. In particular it is measurable.

Theorem 6.2.1 *Let $\phi : \mathbb{D} \to \mathbb{D}$ be a holomorphic function. Then $C_\phi(\mathcal{D}) \subset \mathcal{D}$ if and only if there exists a constant B such that*

$$\int |f'|^2 n_\phi \, dA \leq B \int |f'|^2 \, dA \qquad (f \in \mathcal{D}). \tag{6.3}$$

For the proof, we need a generalization of the change-of-variable formula for non-univalent functions.

Lemma 6.2.2 *Let $\phi : \mathbb{D} \to \mathbb{D}$ be a holomorphic map and let $u : \mathbb{D} \to [0, \infty]$ be a positive measurable function. Then*

$$\int_{\mathbb{D}} (u \circ \phi)|\phi'|^2 \, dA = \int_{\mathbb{D}} u n_\phi \, dA.$$

Proof We may assume that ϕ is non-constant, otherwise the result is obvious. Then $Z := \{z \in \mathbb{D} : \phi'(z) = 0\}$ is a countable set, and $\mathbb{D} \setminus Z$ may be written as a countable union of disjoint rectangles (R_j) on each of which ϕ is a diffeomorphism onto its range. By the usual change-of-variable formula, for each j we have

$$\int_{R_j} (u \circ \phi)|\phi'|^2 \, dA = \int_{\phi(R_j)} u \, dA.$$

Summing over j, we deduce that

$$\int_{\mathbb{D}\setminus Z} (u \circ \phi)|\phi'|^2 \, dA = \int_{\mathbb{D}} u \sum_j 1_{\phi(R_j)} \, dA.$$

Now Z, being countable, is of measure zero. Therefore the left-hand side equals $\int_{\mathbb{D}} (u \circ \phi)|\phi'|^2 \, dA$. Also $\sum_j 1_{\phi(R_j)} = n_\phi$ on $\mathbb{D} \setminus \phi(Z)$, and $\phi(Z)$, being countable, is of measure zero. Thus the right-hand side equals $\int_{\mathbb{D}} u n_\phi \, dA$. Putting together these facts, we obtain the result. □

Proof of Theorem 6.2.1 By the lemma, applied with $u := |f'|^2$, we have

$$\mathcal{D}(f \circ \phi) = \frac{1}{\pi} \int_{\mathbb{D}} |f'|^2 n_\phi \, dA \qquad (f \in \mathcal{D}).$$

Therefore, if (6.3) holds, then $\mathcal{D}(f \circ \phi) < \infty$ for all $f \in \mathcal{D}$, and so $C_\phi(\mathcal{D}) \subset \mathcal{D}$.

Conversely, as already remarked, if $C_\phi(\mathcal{D}) \subset \mathcal{D}$, then $C_\phi : \mathcal{D} \to \mathcal{D}$ is a bounded linear map, and consequently,

$$\|C_\phi(f)\|_{\mathcal{D}} \leq \|C_\phi\| \, \|f\|_{\mathcal{D}} \qquad (f \in \mathcal{D}).$$

This implies that

$$\mathcal{D}(f \circ \phi) \leq 2\|C_\phi\|^2 \mathcal{D}(f) \qquad (f \in \mathcal{D}).$$

By Lemma 6.2.2 again, it follows that

$$\frac{1}{\pi} \int_{\mathbb{D}} |f'|^2 n_\phi \, dA \leq 2\|C_\phi\|^2 \frac{1}{\pi} \int_{\mathbb{D}} |f'|^2 \, dA$$

which shows that (6.3) holds with $B := 2\|C_\phi\|^2$. □

This theorem plays much the same sort of role for composition operators that Theorem 5.1.7 did for multipliers. Just like Theorem 5.1.7, it is unsatisfactory inasmuch as the criterion (6.3) is hard to check in practice. By analogy with the multiplier case, we are therefore led to study measures μ on $\mathbb{D}$ satisfying

$$\int_{\mathbb{D}} |f'|^2\, d\mu \leq \frac{B}{\pi}\int_{\mathbb{D}} |f'|^2\, dA \qquad (f \in \mathcal{D}),$$

or, equivalently,

$$\int_{\mathbb{D}} |h|^2\, d\mu \leq \frac{B}{\pi}\int_{\mathbb{D}} |h|^2\, dA \qquad (h \in \mathcal{B}),$$

where $\mathcal{B}$ is the *Bergman space* on $\mathbb{D}$, defined by

$$\mathcal{B} := \Big\{h \in \mathrm{Hol}(\mathbb{D}) : \|h\|_{\mathcal{B}}^2 := \frac{1}{\pi}\int_{\mathbb{D}} |h|^2\, dA < \infty\Big\}.$$

In other words, using the language of §5.2, we seek to characterize Carleson measures for $\mathcal{B}$. Fortunately, this turns out to be somewhat easier than in the case of Carleson measures for $\mathcal{D}$. We recall from §5.2 that, given an arc $I \subset \mathbb{T}$, the corresponding Carleson box $S(I)$ is defined by

$$S(I) := \{re^{i\theta} : e^{i\theta} \in I,\ 1 - |I| < r < 1\}.$$

Theorem 6.2.3 *Let μ be a finite positive measure on $\mathbb{D}$. The following statements are equivalent.*

(i) *There exists a constant B such that*

$$\int_{\mathbb{D}} |h|^2\, d\mu \leq B\|h\|_{\mathcal{B}}^2 \qquad (h \in \mathcal{B}).$$

(ii) *For $w \in \mathbb{D}$,*

$$\int_{\mathbb{D}} \frac{d\mu(z)}{|1 - \overline{w}z|^4} = O((1 - |w|^2)^{-2}) \qquad (|w| \to 1).$$

(iii) *For arcs $I \subset \mathbb{T}$,*

$$\mu(S(I)) = O(|I|^2) \qquad (|I| \to 0).$$

For the proof, we need an elementary lemma.

Lemma 6.2.4 *If $h \in \mathcal{B}$, then*

$$|h(z)|^2 \leq \frac{3}{\pi}\int_{\mathbb{D}} \frac{(1 - |w|^2)^2}{|1 - \overline{w}z|^4}|h(w)|^2\, dA(w) \qquad (z \in \mathbb{D}). \tag{6.4}$$

Proof For $z \in \mathbb{D}$ and $n \geq 0$, we have

$$\begin{aligned}
&\int_{\mathbb{D}} w^n \frac{(1-|w|^2)^2}{(1-\overline{w}z)^4}\,dA(w) \\
&= \int_{\mathbb{D}} w^n(1-|w|^2)^2 \sum_{k\geq 0}(k+3)(k+2)(k+1)(\overline{w}z)^k\,dA(w) \\
&= \sum_{k\geq 0}(k+3)(k+2)(k+1)z^k \int_0^1 \int_0^{2\pi} r^{n+k}(1-r^2)^2 e^{i(n-k)\theta} r\,d\theta\,dr \\
&= \pi z^n/3.
\end{aligned}$$

By linearity and approximation, it follows that, for all $f \in \mathrm{Hol}(\mathbb{D}) \cap L^1(\mathbb{D}, dA)$,

$$f(z) = \frac{3}{\pi}\int_{\mathbb{D}} f(w)\frac{(1-|w|^2)^2}{(1-\overline{w}z)^4}\,dA(w) \qquad (z \in \mathbb{D}).$$

In particular, this holds for $f = h^2$, where $h \in \mathcal{B}$. Inequality (6.4) follows. □

Proof of Theorem 6.2.3 We first show that (i) implies (ii). Let $h := \psi'$, where $\psi(z) := (z-w)/(1-\overline{w}z)$. A computation gives $|h(z)|^2 = (1-|w|^2)^2/|1-\overline{w}z|^4$ and $\|h\|_{\mathcal{B}}^2 = \mathcal{D}(\psi) = \mathcal{D}(z) = 1$. Therefore, if (i) holds, then

$$\int_{\mathbb{D}} \frac{(1-|w|^2)^2}{|1-\overline{w}z|^4}\,d\mu(z) \leq B \qquad (w \in \mathbb{D}).$$

Thus (ii) holds.

Next we show that (ii) implies (i). For this, we use Lemma 6.2.4. By that lemma, if $h \in \mathcal{B}$, then

$$\begin{aligned}
\int_{\mathbb{D}} |h(z)|^2\,d\mu(z) &\leq \int_{\mathbb{D}}\Big(\frac{3}{\pi}\int_{\mathbb{D}} \frac{(1-|w|^2)^2}{|1-\overline{w}z|^4}|h(w)|^2\,dA(w)\Big)\,d\mu(z) \\
&= \frac{3}{\pi}\int_{\mathbb{D}} |h(w)|^2(1-|w|^2)^2\Big(\int_{\mathbb{D}} \frac{d\mu(z)}{|1-\overline{w}z|^2}\Big)\,dA(w).
\end{aligned}$$

Therefore, if (ii) holds, then there exists a constant B such that

$$\int_{\mathbb{D}} |h|^2\,d\mu \leq \frac{1}{\pi}\int_{\mathbb{D}} B|h|^2\,dA = B\|h\|_{\mathcal{B}}^2 \qquad (h \in \mathcal{B}),$$

and so (i) holds.

Now we show that (ii) implies (iii). Given an arc $I \subset \mathbb{T}$ with $|I| < 1/4$, choose $w \in \mathbb{D}$ so that $|w| = 1 - |I|$ and $w/|w|$ is the midpoint of I. If $z \in S(I)$, then elementary estimates give $|1-\overline{w}z| \leq 3|I|$, and so

$$\int_{\mathbb{D}} \frac{d\mu(z)}{|1-\overline{w}z|^4} \geq \int_{S(I)} \frac{d\mu(z)}{|1-\overline{w}z|^4} \geq \frac{\mu(S(I))}{(3|I|)^4}.$$

Consequently, if (ii) holds, then there exists a constant B such that

$$\frac{\mu(S(I))}{(3|I|)^4} \leq \frac{B}{(1-|w|^2)^2} \leq \frac{B}{|I|^2}.$$

It follows that $\mu(S(I)) \leq 3^4 B|I|^2$, and therefore (iii) holds.

Finally, we prove that (iii) implies (ii). Let $w \in \mathbb{D}$ with $|w| > 1/2$. Using Fubini's theorem, we have

$$\begin{aligned}\int_{\mathbb{D}} \frac{d\mu(z)}{|1-\overline{w}z|^4} &= \int_{\mathbb{D}} \int_{t=|1-\overline{w}z|}^{\infty} 4t^{-5}\,dt\,d\mu(z)\\ &= \int_{t=0}^{\infty} \int_{\{z:|1-\overline{w}z|\leq t\}} 4t^{-5}\,d\mu(z)\,dt\\ &= \int_{t=0}^{\infty} \mu\{z \in \mathbb{D} : |1-\overline{w}z| \leq t\} 4t^{-5}\,dt.\end{aligned}$$

Now $\{z \in \mathbb{D} : |1-\overline{w}z| \leq t\} = \{z \in \mathbb{D} : |z - 1/\overline{w}| \leq t/|w|\}$. This set is contained in $S(I)$ for some arc I with $|I| = 8t$, and it is even empty if $t < 1 - |w|$. Thus, if (iii) holds, then

$$\mu\{z \in \mathbb{D} : |1-\overline{w}z| \leq t\} \leq \begin{cases} 0, & t < 1-|w|,\\ Ct^2, & t \geq 1-|w|,\end{cases}$$

where C is a constant independent of w. Feeding this into the previous estimate, we obtain

$$\int_{\mathbb{D}} \frac{d\mu(z)}{|1-\overline{w}z|^4} \leq \int_{t=1-|w|}^{\infty} Ct^2 4t^{-5}\,dt = \frac{2C}{(1-|w|)^2} \leq \frac{8C}{(1-|w|^2)^2}.$$

This proves that (ii) holds. □

We can now read off the characterization of those ϕ for which $C_\phi(\mathcal{D}) \subset \mathcal{D}$.

Theorem 6.2.5 *Let $\phi : \mathbb{D} \to \mathbb{D}$ be a holomorphic map such that $\phi \in \mathcal{D}$. The following statements are equivalent.*

(i) $C_\phi(\mathcal{D}) \subset \mathcal{D}$.

(ii) *For $w \in \mathbb{D}$,*

$$\frac{1}{\pi}\int_{\mathbb{D}} \frac{n_\phi(z)}{|1-\overline{w}z|^4}\,dA(z) = O((1-|w|^2)^{-2}) \qquad (|w| \to 1). \tag{6.5}$$

(iii) *For arcs $I \subset \mathbb{T}$,*

$$\frac{1}{\pi}\int_{S(I)} n_\phi\,dA = O(|I|^2) \qquad (|I| \to 0).$$

Proof By Lemma 6.2.2, the fact that $\phi \in \mathcal{D}$ implies that $d\mu := n_\phi \, dA$ is a finite measure on $\mathbb{D}$. The result is therefore an immediate consequence of Theorems 6.2.1 and 6.2.3 combined. □

As an application of this theorem, we derive a simple sufficient condition for $C_\phi(\mathcal{D}) \subset \mathcal{D}$. It is part (ii) of the following result.

Theorem 6.2.6 *Let $\phi : \mathbb{D} \to \mathbb{D}$ be a holomorphic map such that $\phi^k \in \mathcal{D}$ for all $k \geq 1$.*

(i) *If $C_\phi(\mathcal{D}) \subset \mathcal{D}$, then $\mathcal{D}(\phi^k) = O(k)$ as $k \to \infty$.*
(ii) *If $\mathcal{D}(\phi^k) = O(1)$ as $k \to \infty$, then $C_\phi(\mathcal{D}) \subset \mathcal{D}$.*

Proof (i) If $C_\phi(\mathcal{D}) \subset \mathcal{D}$, then

$$\mathcal{D}(\phi^k) \leq \|\phi^k\|_{\mathcal{D}}^2 = \|C_\phi(z^k)\|_{\mathcal{D}}^2 \leq \|C_\phi\|^2 \|z^k\|_{\mathcal{D}}^2 = \|C_\phi\|^2 (k+1) \qquad (k \geq 0).$$

(ii) Assume now that $\mathcal{D}(\phi^k) \leq B$ for all $k \geq 1$. We shall verify that (6.5) holds. For each $w \in \mathbb{D}$, we have

$$\begin{aligned}
\frac{1}{\pi}\int_{\mathbb{D}} \frac{n_\phi(z)}{|1-\overline{w}z|^4}\, dA(z) &\leq \frac{2^4}{\pi}\int_{\mathbb{D}} \frac{n_\phi(z)}{|1-\overline{w}^2 z^2|^4}\, dA(z) \\
&= \frac{2^4}{\pi}\int_{\mathbb{D}} \sum_{k\geq 1} (k+2)(k+1)k|\overline{w}^2 z^2|^{k-1} n_\phi(z)\, dA(z) \\
&\leq 6.2^4 \sum_{k\geq 1} k|w^2|^{k-1} \frac{1}{\pi}\int_{\mathbb{D}} k^2 |z|^{2k-2} n_\phi(z)\, dA(z) \\
&= 6.2^4 \sum_{k\geq 1} k|w^2|^{k-1} \mathcal{D}(\phi^k). \\
&\leq \frac{6.2^4 B}{(1-|w|^2)^2}.
\end{aligned}$$

Hence (6.5) holds, as required. □

In Theorem 5.1.4 we saw that the multiplication operators could be characterized as those operators on $\mathcal{D}$ whose adjoints have the reproducing kernels as eigenvectors. The following result is the analogue for composition operators.

Theorem 6.2.7 *If $\phi : \mathbb{D} \to \mathbb{D}$ is holomorphic and $C_\phi(\mathcal{D}) \subset \mathcal{D}$, then*

$$C_\phi^*(k_w) = k_{\phi(w)} \qquad (w \in \mathbb{D}).$$

Conversely, if $T : \mathcal{D} \to \mathcal{D}$ is a bounded operator such that T^* maps each reproducing kernel to a reproducing kernel, then there exists a holomorphic map $\phi : \mathbb{D} \to \mathbb{D}$ such that $T = C_\phi$.

Proof Assume that $C_\phi(\mathcal{D}) \subset \mathcal{D}$. Let $w \in \mathbb{D}$. For each $f \in \mathcal{D}$, we have

$$\langle f, C_\phi^*(k_w)\rangle_{\mathcal{D}} = \langle C_\phi(f), k_w\rangle_{\mathcal{D}} = f(\phi(w)) = \langle f, k_{\phi(w)}\rangle_{\mathcal{D}}.$$

It follows that $C_\phi^*(k_w) = k_{\phi(w)}$.

For the converse, suppose that, for each $w \in \mathbb{D}$, we have $T^*(k_w) = k_{\phi(w)}$, where $\phi : \mathbb{D} \to \mathbb{D}$ is some unknown function. For each $f \in \mathcal{D}$, we then have

$$(Tf)(w) = \langle T(f), k_w\rangle_{\mathcal{D}} = \langle f, T^*(k_w)\rangle_{\mathcal{D}} = \langle f, k_{\phi(w)}\rangle_{\mathcal{D}} = f(\phi(w)) \qquad (w \in \mathbb{D}).$$

In particular, taking $f(z) = z$, we see that $\phi = T(z) \in \mathrm{Hol}(\mathbb{D})$. The calculation above then shows that $T(f) = C_\phi(f)$ for all $f \in \mathcal{D}$, whence the result. □

Corollary 6.2.8 *If $\phi : \mathbb{D} \to \mathbb{D}$ is holomorphic and $C_\phi(\mathcal{D}) \subset \mathcal{D}$, then*

$$\log\Bigl(\frac{1}{1-|\phi(z)|^2}\Bigr) = O\Bigl(\log\Bigl(\frac{1}{1-|z|^2}\Bigr)\Bigr) \qquad (|z| \to 1).$$

Proof From the theorem, $\|k_{\phi(z)}\|_{\mathcal{D}}^2 = \|C_\phi^*(k_z)\|_{\mathcal{D}}^2 \le \|C_\phi\|^2\|k_z\|_{\mathcal{D}}^2$ $(z \in \mathbb{D})$. Recalling that $\|k_w\|_{\mathcal{D}}^2 = k_w(w) = |w|^{-2}\log(1/(1-|w|^2))$, we obtain the result. □

Exercises 6.2

1. Let $\phi : \mathbb{D} \to \mathbb{D}$ be a non-constant holomorphic map. Show that its counting function n_ϕ is lower semicontinuous on $\mathbb{D}$.
2. Let μ be a finite positive measure on $\mathbb{D}$. By examining the proof of Theorem 6.2.3, prove that the following 'little o' statements are equivalent.

   (i) For $w \in \mathbb{D}$,

   $$\int_{\mathbb{D}} \frac{d\mu(z)}{|1-\overline{w}z|^4} = o((1-|w|^2)^{-2}) \qquad (|w| \to 1).$$

   (ii) For arcs $I \subset \mathbb{T}$,

   $$\mu(S(I)) = o(|I|^2) \qquad (|I| \to 0).$$

6.3 Compactness criteria

Now that we have characterized those ϕ for which $C_\phi : \mathcal{D} \to \mathcal{D}$ is a bounded operator, it seems only natural to ask when C_ϕ is compact. An answer is provided by the following analogue of Theorem 6.2.5, in which the 'big O' conditions are replaced by corresponding 'little o' ones.

Theorem 6.3.1 *Let $\phi : \mathbb{D} \to \mathbb{D}$ be a holomorphic map such that $\phi \in \mathcal{D}$. The following statements are equivalent.*

(i) $C_\phi : \mathcal{D} \to \mathcal{D}$ *is compact.*

(ii) *For* $w \in \mathbb{D}$,

$$\frac{1}{\pi}\int_{\mathbb{D}} \frac{n_\phi(z)}{|1-\overline{w}z|^4}\,dA(z) = o((1-|w|^2)^{-2}) \qquad (|w| \to 1).$$

(iii) *For arcs* $I \subset \mathbb{T}$,

$$\frac{1}{\pi}\int_{S(I)} n_\phi \,dA = o(|I|^2) \qquad (|I| \to 0).$$

Proof The equivalence between (ii) and (iii) follows from Exercise 6.2.2. We shall prove the equivalence between (i) and (ii).

Suppose first that (i) holds. For $w \in \mathbb{D} \setminus \{0\}$, set

$$\psi_w(z) := \frac{|w|}{w}\frac{w-z}{1-\overline{w}z} \qquad (z \in \mathbb{D}).$$

Then $\|\psi_w\|_{\mathcal{D}}^2 = \|\psi_w\|_{H^2}^2 + \mathcal{D}(\psi_w) = 2$ for all $w \in \mathbb{D}\setminus\{0\}$, and $\psi_w \to 1$ pointwise as $|w| \to 1$. Hence $\psi_w \to 1$ weakly in $\mathcal{D}$ as $|w| \to 1$. As C_ϕ is compact, it follows that $C_\phi(\psi_w) \to C_\phi(1)$ in norm as $|w| \to 1$, in other words, that $\|\psi_w \circ \phi - 1\|_{\mathcal{D}} \to 0$ as $|w| \to 1$. This implies that $\mathcal{D}(\psi_w \circ \phi) \to 0$ as $|w| \to 1$. Also, by Lemma 6.2.2,

$$\mathcal{D}(\psi_w \circ \phi) = \frac{1}{\pi}\int_{\mathbb{D}} \frac{(1-|w|^2)^2}{|1-\overline{w}z|^4} n_\phi(z)\,dA(z) \qquad (w \in \mathbb{D} \setminus \{0\}).$$

Putting these facts together, we deduce that (ii) holds.

Now suppose that (ii) holds. Let (f_k) be a sequence such that $f_k \to 0$ weakly in $\mathcal{D}$. We need to show that $C_\phi(f_k) \to 0$ in norm. Since

$$\|C_\phi(f_k)\|_{\mathcal{D}}^2 \le |f_k(\phi(0))|^2 + 2\mathcal{D}(f_k \circ \phi)$$

and $f_k \to 0$ pointwise on $\mathbb{D}$, it suffices to show that $\mathcal{D}(f_k \circ \phi) \to 0$ as $k \to \infty$. We now estimate this quantity. Using Lemmas 6.2.2 and 6.2.4, we have

$$\begin{aligned}
\mathcal{D}(f_k \circ \phi) &= \frac{1}{\pi}\int_{\mathbb{D}} |f_k'(z)|^2 n_\phi(z)\,dA(z) \\
&\le \frac{1}{\pi}\int_{\mathbb{D}} \Big(\frac{3}{\pi}\int_{\mathbb{D}} \frac{(1-|w|^2)^2}{|1-\overline{w}z|^4}|f_k'(w)|^2\,dA(w)\Big) n_\phi(z)\,dA(z) \\
&= \frac{3}{\pi}\int_{\mathbb{D}} \Big(\frac{1}{\pi}\int_{\mathbb{D}} \frac{(1-|w|^2)^2}{|1-\overline{w}z|^4} n_\phi(z)\,dA(z)\Big)|f_k'(w)|^2\,dA(w).
\end{aligned}$$

Since (ii) holds, given $\epsilon > 0$, there exists $r \in (0,1)$ such that

$$\frac{1}{\pi}\int_{\mathbb{D}} \frac{(1-|w|^2)^2}{|1-\overline{w}z|^4} n_\phi(z)\,dA(z) < \epsilon \qquad (r < |w| < 1).$$

Also, it is clear that, if $|w| \le r$, then

$$\frac{1}{\pi}\int_{\mathbb{D}} \frac{(1-|w|^2)^2}{|1-\overline{w}z|^4} n_\phi(z)\,dA(z) \le \frac{1}{\pi}\int_{\mathbb{D}} \frac{n_\phi(z)}{(1-r)^4}\,dA(z) = \frac{\mathcal{D}(\phi)}{(1-r)^4}.$$

Substituting these estimates into the inequality for $\mathcal{D}(f_k \circ \phi)$, we obtain

$$\mathcal{D}(f_k \circ \phi) \le \frac{3}{\pi}\int_{r<|w|<1} \epsilon |f_k'(w)|^2\,dA(w) + \frac{3}{\pi}\int_{|w|\le r} \frac{\mathcal{D}(\phi)}{(1-r)^4}|f_k'(w)|^2\,dA(w).$$

The fact that $f_k \to 0$ weakly in $\mathcal{D}$ implies both that $\mathcal{D}(f_k)$ remains bounded and that $f_k \to 0$ locally uniformly on $\mathbb{D}$. Therefore

$$\limsup_{k\to\infty} \mathcal{D}(f_k \circ \phi) \le \epsilon \sup_k \mathcal{D}(f_k).$$

As ϵ is arbitrary, we deduce that $\lim_{k\to\infty} \mathcal{D}(f_k \circ \phi) = 0$, as desired. □

Theorem 6.2.6 and Corollary 6.2.8 also have 'little o' versions for compact operators. These are outlined in Exercises 6.3.1 and 6.3.2 respectively.

We next characterize those ϕ for which C_ϕ is a Hilbert–Schmidt operator. Recall that a linear map T from Hilbert space $\mathcal{H}$ to itself is called a *Hilbert–Schmidt operator* if there is an orthonormal basis (e_k) of $\mathcal{H}$ such that $\sum_k \|Te_k\|^2 < \infty$. It can be shown that, if this sum is finite for one orthonormal basis, then it is finite for all of them, and its value is always the same. Also, every Hilbert–Schmidt operator is compact, but not conversely.

Theorem 6.3.2 *Let $\phi : \mathbb{D} \to \mathbb{D}$ be a holomorphic map. The following statements are equivalent.*

(i) *$C_\phi : \mathcal{D} \to \mathcal{D}$ is a Hilbert–Schmidt operator.*

(ii) $\displaystyle\sum_{k\ge1} \frac{\mathcal{D}(\phi^k)}{k+1} < \infty.$

(iii) $\displaystyle\frac{1}{\pi}\int_{\mathbb{D}} \frac{n_\phi(z)}{(1-|z|^2)^2}\,dA(z) < \infty.$

Proof We first prove the equivalence of (i) and (ii). Set $e_k(z) := z^k/\sqrt{k+1}$. Then $(e_k)_{k\ge0}$ is an orthonormal basis of $\mathcal{D}$, so C_ϕ is Hilbert–Schmidt if and only if $\sum_k \|C_\phi(e_k)\|_{\mathcal{D}}^2 < \infty$. Now $C_\phi(e_k) = \phi^k/\sqrt{k+1}$, so

$$\frac{\mathcal{D}(\phi^k)}{k+1} \le \|C_\phi(e_k)\|_{\mathcal{D}}^2 \le \frac{|\phi(0)|^{2k}}{k+1} + 2\frac{\mathcal{D}(\phi^k)}{k+1} \qquad (k \ge 0).$$

Also, we clearly have $\sum_k |\phi(0)|^{2k}/(k+1) < \infty$. Therefore C_ϕ is Hilbert–Schmidt if and only if $\sum_k \mathcal{D}(\phi^k)/(k+1) < \infty$.

Now we establish the equivalence of (ii) and (iii). By Lemma 6.2.2, we have

$$\frac{\mathcal{D}(\phi^k)}{k+1} \asymp \frac{1}{\pi}\int_{\mathbb{D}} k|z|^{2k-2} n_\phi(z)\, dA(z) \qquad (k \geq 1),$$

and

$$\sum_{k\geq 1} \frac{1}{\pi}\int_{\mathbb{D}} k|z|^{2k-2} n_\phi(z)\, dA(z) = \frac{1}{\pi}\int_{\mathbb{D}} \frac{n_\phi(z)}{(1-|z|^2)^2}\, dA(z).$$

Combining these observations, we see that (ii) and (iii) are indeed equivalent. □

Corollary 6.3.3 *If $\phi \in \mathcal{D} \cap H^\infty$ with $\|\phi\|_{H^\infty} < 1$, then $C_\phi : \mathcal{D} \to \mathcal{D}$ is Hilbert–Schmidt.*

Proof For each $k \geq 1$, we have $\mathcal{D}(\phi^k) \leq k^2 \|\phi\|_{H^\infty}^{2k-2} \mathcal{D}(\phi)$. Since $\|\phi\|_{H^\infty} < 1$, it follows that $\sum_k \mathcal{D}(\phi^k)/(k+1) < \infty$. □

In the other direction, it would be too much to expect that $\|\phi\|_{H^\infty} < 1$ whenever C_ϕ is Hilbert–Schmidt. Nonetheless, the following theorem shows that there is a sense in which the radial limit $|\phi^*|$ cannot equal 1 on too much of the unit circle. We recall that a property holds quasi-everywhere (q.e.) on $\mathbb{T}$ if it holds everywhere on $\mathbb{T}$ outside a set of outer logarithmic capacity zero.

Theorem 6.3.4 *If $\phi : \mathbb{D} \to \mathbb{D}$ is a holomorphic map such that $C_\phi : \mathcal{D} \to \mathcal{D}$ is Hilbert–Schmidt, then $|\phi^*| < 1$ q.e. on $\mathbb{T}$.*

Proof First of all, since C_ϕ maps $\mathcal{D}$ into itself, we have $\phi \in \mathcal{D}$. Therefore the radial limit ϕ^* exists q.e., and by the weak-type inequality for capacity Theorem 3.3.1,

$$c^*(|\phi^*| \geq 1) \leq \inf_{t<1} c^*(|\phi^*| > t) \leq \inf_{t<1} A\|\phi\|_{\mathcal{D}}^2/t^2 = A\|\phi\|_{\mathcal{D}}^2,$$

where A is an absolute constant. Further, for each k we have $\phi^k = C_\phi(z^k) \in \mathcal{D}$ and so, repeating the argument with ϕ replaced by ϕ^k, we get

$$c^*(|\phi^*| \geq 1) \leq A\|\phi^k\|_{\mathcal{D}}^2 \qquad (k \geq 1). \tag{6.6}$$

On the other hand, as C_ϕ is Hilbert–Schmidt, we have

$$\sum_{k\geq 0} \frac{\|\phi^k\|_{\mathcal{D}}^2}{k+1} = \sum_{k\geq 0} \|C_\phi(e_k)\|_{\mathcal{D}}^2 < \infty,$$

where once again $(e_k)_{k\geq 0}$ denotes the orthonormal basis $e_k(z) := z^k/\sqrt{k+1}$. In particular, we must have $\liminf_{k\to\infty} \|\phi^k\|_{\mathcal{D}} = 0$. Combining this with (6.6), we deduce that $c^*(|\phi^*| \geq 1) = 0$, whence the result. □

There is an analogous theorem in the case when C_ϕ is merely compact, but it is then necessary to replace logarithmic capacity by Riesz capacity. An outline is sketched in Exercise 6.3.3 below.

Exercises 6.3

1. Let $\phi : \mathbb{D} \to \mathbb{D}$ be a holomorphic map such that $\phi^k \in \mathcal{D}$ for all $k \geq 1$.
   (i) Show that, if $C_\phi : \mathcal{D} \to \mathcal{D}$ is compact, then $\mathcal{D}(\phi^k) = o(k)$ as $k \to \infty$. [Hint: Show that $z^k/\sqrt{k+1} \to 0$ weakly in $\mathcal{D}$ as $k \to \infty$.]
   (ii) Show that, if $\mathcal{D}(\phi^k) = o(1)$ as $k \to \infty$, then $C_\phi : \mathcal{D} \to \mathcal{D}$ is compact.
2. Let $\phi : \mathbb{D} \to \mathbb{D}$ be a holomorphic map such that $C_\phi : \mathcal{D} \to \mathcal{D}$ is compact. Show that
$$\log\Big(\frac{1}{1-|\phi(z)|^2}\Big) = o\Big(\log\Big(\frac{1}{1-|z|^2}\Big)\Big) \qquad (|z| \to 1).$$
   [Hint: Show that $k_w/\|k_w\|_{\mathcal{D}} \to 0$ weakly in $\mathcal{D}$ as $|w| \to 1$.]
3. Let $\phi : \mathbb{D} \to \mathbb{D}$ be a holomorphic map such that $C_\phi : \mathcal{D} \to \mathcal{D}$ is compact.
   (i) Show that, for all $\alpha \in (0,1)$,
$$\frac{1}{\pi}\int_{\mathbb{D}} \frac{|\phi'(z)|^2(1-|z|^2)^\alpha}{(1-|\phi(z)|^2)^2}\,dA(z) < \infty.$$
   [Hint: Use the result of the preceding exercise.]
   (ii) Deduce that, for all $\alpha \in (0,1)$,
$$\sum_{k\geq 0} \frac{\mathcal{D}_\alpha(\phi^k)}{k+1} < \infty.$$
   (iii) Deduce that $\liminf_{k\to\infty} \|\phi^k\|_{\mathcal{D}_\alpha} = 0$ for all $\alpha \in (0,1)$.
   (iv) Using the weak-type inequality for the Riesz capacity c_α established in Exercise 3.3.4, conclude that $c_\alpha^*(|\phi^*| \geq 1) = 0$ for all $\alpha \in (0,1)$.
4. Let $\alpha > 1/2$ and let ϕ_α be a conformal mapping of the unit disk onto the domain $\{x+iy : 0 < x < 1,\ |y| < ((1-x^2)/2)^\alpha\}$. For which values of α is $C_{\phi_\alpha} : \mathcal{D} \to \mathcal{D}$ compact?

Notes on Chapter 6

§6.1

Theorem 6.1.1 is due to Arazy and Fisher [8], though our formulation of the result is a little different. It is possible to weaken the assumption (i) to a certain extent and still conclude that $\mathcal{H} = \mathcal{D}$. For more on this we refer to [8].

§6.2 and §6.3

The subject of composition operators on holomorphic function spaces is vast. For background on this topic, we refer to the books of Cowen and MacCluer [36] and of Shapiro [113].

The idea of exploiting counting functions in the context of composition operators is due to Shapiro [112]. He used them to characterize those ϕ for which C_ϕ is a compact operator on H^2. Theorems 6.2.5 and 6.3.1 are due to MacCluer and Shapiro [73]. The main ideas behind these theorems are also present in the article of Luecking [72], and these were recently exploited by Lefèvre, Li, Queffélec and Rodríguez-Piazza [71] to establish criteria for C_ϕ to belong to the Schatten classes. The characterization of Carleson measures for the Bergman space, Theorem 6.2.3, is due to Hastings [56]. Theorem 6.2.6 is taken from [45].

Theorem 6.3.4 is due to Gallardo-Gutiérrez and González [49], who also gave an example of a function ϕ such that C_ϕ is compact but $c^*(|\phi^*| \geq 1) > 0$. There is also a quantitative version of their result, proved in [45]: if C_ϕ is Hilbert–Schmidt, then

$$\int_0^1 \frac{c^*(|\phi^*| > t)}{1-t} \log\Big(\frac{1}{1-t}\Big)\, dt < \infty,$$

and this is, in a certain sense, best possible. The article [71] contains further results along these lines.

7

Harmonically weighted Dirichlet spaces

In §1.6, we introduced the notion of a weighted Dirichlet integral. Given a measurable function $w : \mathbb{D} \to (0, \infty)$, we defined

$$\mathcal{D}_w(f) := \frac{1}{\pi} \int_{\mathbb{D}} |f'(z)|^2 w(z)\, dA(z) \quad (f \in \operatorname{Hol}(\mathbb{D})),$$
$$\mathcal{D}_w := \{f \in \operatorname{Hol}(\mathbb{D}) : \mathcal{D}_w(f) < \infty\}.$$

In this chapter we are going to examine in some detail a special class of weights, namely those w that are harmonic functions on $\mathbb{D}$. There are (at least) two reasons why these particular weights are important.

Firstly, a general harmonic weight w can be synthesized as an average of extreme harmonic weights, which turn out to be particularly simple to treat. In particular, this is true of the constant weight $w \equiv 1$, which corresponds to the standard Dirichlet integral. We shall see in the course of the chapter that this idea is a very useful tool.

Secondly, the harmonically weighted spaces $\mathcal{D}_w$ play a central role in the theory of shift-invariant subspaces of the Dirichlet space. This will be the subject of the next chapter.

7.1 $\mathcal{D}_\mu$-spaces and the local Dirichlet integral

It is well known that every positive harmonic function w on $\mathbb{D}$ can be represented as the Poisson integral $P\mu$ of a finite (positive, Borel) measure μ on $\mathbb{T}$, namely

$$w(z) = P\mu(z) := \int_{\mathbb{T}} \frac{1 - |z|^2}{|z - \zeta|^2}\, d\mu(\zeta) \qquad (z \in \mathbb{D}).$$

This leads us to make the following definition.

Definition 7.1.1 Let μ be a finite measure on $\mathbb{T}$ with $\mu(\mathbb{T}) > 0$. The associated *harmonically weighted Dirichlet integral* and *harmonically weighted Dirichlet space* are defined respectively by

$$\mathcal{D}_\mu(f) := \frac{1}{\pi}\int_{\mathbb{D}} |f'(z)|^2 P\mu(z)\, dA(z) \quad (f \in \mathrm{Hol}(\mathbb{D})),$$
$$\mathcal{D}_\mu := \{f \in \mathrm{Hol}(\mathbb{D}) : \mathcal{D}_\mu(f) < \infty\}.$$

Clearly $\mathcal{D}_\mu$ is a subspace of $\mathrm{Hol}(\mathbb{D})$ containing the polynomials. The following result tells us a bit more.

Theorem 7.1.2 *Let μ be a finite measure on $\mathbb{T}$ with $\mu(\mathbb{T}) > 0$. Then $\mathcal{D}_\mu \subset H^2$, and $\mathcal{D}_\mu$ is a Hilbert space with respect to the norm $\|\cdot\|_{\mathcal{D}_\mu}$ given by*

$$\|f\|^2_{\mathcal{D}_\mu} := \|f\|^2_{H^2} + \mathcal{D}_\mu(f) \qquad (f \in \mathcal{D}_\mu).$$

Proof This is actually a special case of Theorem 1.6.3. All we have to show is that $\liminf_{|z|\to 1} P\mu(z)/(1-|z|) > 0$. And this is indeed true, because

$$\frac{P\mu(z)}{1-|z|} = \int_{\mathbb{T}} \frac{1+|z|}{|z-\zeta|^2}\, d\mu(\zeta) \geq \frac{\mu(\mathbb{T})}{4} > 0 \quad (z \in \mathbb{D}).$$ □

There are three special cases particularly worthy of note.

The first is when $\mu = m$, normalized Lebesgue measure on $\mathbb{T}$. In this case we have $Pm \equiv 1$, so $\mathcal{D}_m(f) = \mathcal{D}(f)$, the standard Dirichlet integral.

The second case is when $\mu = 0$. This is not actually covered by the definition, but, in view of the last result, it is logical to extend Definition 7.1.1 to this case by defining $\mathcal{D}_0(f) := 0$ and $\mathcal{D}_0 := H^2$.

Finally, we consider the case when $\mu = \delta_\zeta$, the Dirac measure at $\zeta \in \mathbb{T}$. In this case, we write $\mathcal{D}_\zeta(f)$ in place of $\mathcal{D}_{\delta_\zeta}(f)$. We thus have

$$\mathcal{D}_\zeta(f) = \frac{1}{\pi}\int_{\mathbb{D}} |f'(z)|^2 \frac{1-|z|^2}{|z-\zeta|^2}\, dA(z).$$

This is called the *local Dirichlet integral* of f at ζ, and the corresponding space $\mathcal{D}_\zeta$ is called the *local Dirichlet space* at ζ.

Theorem 7.1.3 *Let μ be a finite measure on $\mathbb{T}$. Then*

$$\mathcal{D}_\mu(f) = \int_{\mathbb{T}} \mathcal{D}_\zeta(f)\, d\mu(\zeta) \qquad (f \in \mathrm{Hol}(\mathbb{D})). \tag{7.1}$$

In particular,

$$\mathcal{D}(f) = \frac{1}{2\pi}\int_{\mathbb{T}} \mathcal{D}_\zeta(f)\, |d\zeta| \qquad (f \in \mathrm{Hol}(\mathbb{D})). \tag{7.2}$$

Proof By Fubini's theorem:

$$\begin{aligned}\mathcal{D}_\mu(f) &= \frac{1}{\pi}\int_{\mathbb{D}} |f'(z)|^2\Big(\int_{\mathbb{T}} \frac{1-|z|^2}{|z-\zeta|^2}\,d\mu(\zeta)\Big)\,dA(z)\\ &= \int_{\mathbb{T}}\Big(\frac{1}{\pi}\int_{\mathbb{D}} |f'(z)|^2\frac{1-|z|^2}{|z-\zeta|^2}\,dA(z)\Big)\,d\mu(\zeta)\\ &= \int_{\mathbb{T}} \mathcal{D}_\zeta(f)\,d\mu(\zeta).\end{aligned}$$

This gives (7.1), and (7.2) follows by taking μ to be Lebesgue measure. □

Thanks to this useful result, it frequently suffices to treat $\mathcal{D}_\zeta(f)$ rather than $\mathcal{D}_\mu(f)$ for general μ. This can help simplify the notation.

Exercises 7.1

1. Verify directly that $\mathcal{D}_\mu(z^n) = n\mu(\mathbb{T})$.
2. Let $f \in \mathrm{Hol}(\mathbb{D})$ and suppose that f' is bounded on $\mathbb{D}$. Show that $f \in \mathcal{D}_\mu$ for every finite measure μ on $\mathbb{T}$, and that

$$\mathcal{D}_\mu(f) \le \|f'\|_\infty^2\mu(\mathbb{T}).$$

3. Let $(f_n), f \in \mathrm{Hol}(\mathbb{D})$, and suppose that $f_n \to f$ locally uniformly on $\mathbb{D}$. Show that, for each finite measure μ on $\mathbb{T}$,

$$\mathcal{D}_\mu(f) \le \liminf_{n\to\infty} \mathcal{D}_\mu(f_n).$$

4. Let μ be a probability measure on $\mathbb{T}$.
   (i) Given $f \in \mathrm{Hol}(\mathbb{D})$ and $\lambda \in \mathbb{T}$, define $f_\lambda(z) := f(\lambda^{-1}z)$. Show that

$$\frac{1}{2\pi}\int_{\mathbb{T}} \mathcal{D}_\mu(f_\lambda)\,|d\lambda| = \mathcal{D}(f).$$

   Deduce that, if $f \in \mathcal{D}$, then $f_\lambda \in \mathcal{D}_\mu$ for at least one $\lambda \in \mathbb{T}$.
   (ii) Deduce that $\mathcal{D}_\mu \not\subset H^\infty$.
   (iii) Deduce that $\mathcal{D}_\mu$ is not an algebra.

7.2 The local Douglas formula

Recall that, given $f \in \mathrm{Hol}(\mathbb{D})$ and $\zeta \in \mathbb{T}$, we write $f^*(\zeta)$ for the radial limit $\lim_{r\to 1^-} f(r\zeta)$, whenever it exists. In §1.5, we established a formula of Douglas expressing $\mathcal{D}(f)$ in terms of f^*. We now seek to extend this formula to the local Dirichlet integral $\mathcal{D}_\zeta(f)$. As a by-product, we shall also derive a formula for $\mathcal{D}_\zeta(f)$ in terms of the Taylor coefficients of f, analogous to Theorem 1.1.2. Both are consequences of the following basic theorem, which establishes a criterion for membership in the local Dirichlet space $\mathcal{D}_\zeta$.

Theorem 7.2.1 *Let $f \in \mathrm{Hol}(\mathbb{D})$ and let $\zeta \in \mathbb{T}$. Then $\mathcal{D}_\zeta(f) < \infty$ if and only if $f(z) = a + (z - \zeta)g(z)$, where $g \in H^2$ and $a \in \mathbb{C}$. In this case $\mathcal{D}_\zeta(f) = \|g\|_{H^2}^2$, and $f^*(\zeta) = a$. Moreover, $f(z) \to f^*(\zeta)$ as $z \to \zeta$ in each oricyclic approach region $|z - \zeta| < \kappa(1 - |z|^2)^{1/2}$ $(\kappa > 0)$.*

For the proof, we need two lemmas.

Lemma 7.2.2 *For $n \geq 1$, define $p_n(z) := nz^n - \sum_{k=0}^{n-1} z^k$. Then $(p_n)_{n\geq1}$ is an orthogonal basis of H^2, and $\|p_n\|_{H^2}^2 = n(n+1)$ for all n.*

Proof Let $1 \leq m < n$. Writing $p_m(z) = \sum_{k=0}^{m-1}(z^m - z^k)$, we have

$$\langle p_n, p_m\rangle_{H^2} = \sum_{k=0}^{m-1}\Big(\langle p_n, z^m\rangle_{H^2} - \langle p_n, z^k\rangle_{H^2}\Big) = \sum_{k=0}^{m-1}(-1+1) = 0.$$

Thus $(p_n)_{n\geq1}$ is an orthogonal sequence in H^2.

If $f \in H^2$ and $f \perp (p_n)_{n\geq1}$, then (writing $p_0 := 0$) we have $p_n - p_{n-1} = n(z^n - z^{n-1})$, and so

$$\langle f, z^n\rangle_{H^2} - \langle f, z^{n-1}\rangle_{H^2} = \langle f, p_n - p_{n-1}\rangle_{H^2}/n = 0 \qquad (n \geq 1).$$

Therefore all the Taylor coefficients of f are equal, which, together with the fact that $f \in H^2$, implies that $f = 0$. Thus $(p_n)_{n\geq1}$ is an orthogonal basis of H^2.

Finally, for each n we have $\|p_n\|_{H^2}^2 = n^2 + \sum_{k=0}^{n-1} 1^2 = n(n+1)$. □

To state the second lemma, we introduce the space

$$\mathcal{A} := \Big\{h \in \mathrm{Hol}(\mathbb{D}) : \|h\|_{\mathcal{A}}^2 := \frac{1}{\pi}\int_{\mathbb{D}} |h(z)|^2(1 - |z|^2)\, dA(z) < \infty\Big\}. \tag{7.3}$$

Note that $\mathcal{A}$ is a Hilbert space. The monomials $(z^n)_{n\geq0}$ form an orthogonal basis of $\mathcal{A}$, and $\|z^n\|_{\mathcal{A}}^2 = 1/(n+1)(n+2)$ for all n. Also, convergence in $\mathcal{A}$ implies local uniform convergence in $\mathbb{D}$ (see Exercise 7.2.1).

Lemma 7.2.3 *Define $T : \mathrm{Hol}(\mathbb{D}) \to \mathrm{Hol}(\mathbb{D})$ by*

$$Tg(z) := \frac{((z-1)g(z))'}{z-1} \qquad (g \in \mathrm{Hol}(\mathbb{D})).$$

Then the restriction of T to H^2 maps H^2 isometrically onto $\mathcal{A}$.

Proof Let $(p_n)_{n\geq1}$ be the sequence defined in Lemma 7.2.2. A simple calculation shows that $(z-1)p_n(z) = nz^{n+1} - (n+1)z^n + 1$, and thus $Tp_n(z) = n(n+1)z^{n-1}$.

Therefore $(Tp_n)_{n\geq 1}$ is an orthogonal sequence in $\mathcal{A}$. Furthermore, for each $n \geq 1$,

$$\|Tp_n\|_{\mathcal{A}}^2 = n^2(n+1)^2\|z^{n-1}\|_{\mathcal{A}}^2 = n(n+1) = \|p_n\|_{H^2}^2.$$

It follows that $\|Tg\|_{\mathcal{A}} = \|g\|_{H^2}$ whenever g is a finite linear combination of (p_n).

For a general $g \in H^2$, we argue as follows. Since $(p_n)_{n\geq 1}$ spans a dense subspace of H^2, we can find a sequence (g_k) in H^2 such that each g_k is a finite linear combination of p_n and $\|g_k - g\|_{H^2} \to 0$. By what we have already proved, $\|Tg_k - Tg_l\|_{\mathcal{A}} = \|g_k - g_l\|_{H^2}$ for all k, l, so (Tg_k) is a Cauchy sequence in $\mathcal{A}$. As $\mathcal{A}$ is complete, there exists $h \in \mathcal{A}$ such that $\|Tg_k - h\|_{\mathcal{A}} \to 0$. Then both $g_k \to g$ and $Tg_k \to h$ locally uniformly in $\mathbb{D}$, so the continuity properties of T imply that $h = Tg$. Thus

$$\|Tg\|_{\mathcal{A}} = \lim_{k\to\infty} \|Tg_k\|_{\mathcal{A}} = \lim_{k\to\infty} \|g_k\|_{H^2} = \|g\|_{H^2}.$$

We conclude that T is an isometry from H^2 into $\mathcal{A}$.

Finally, since the image of T is closed in $\mathcal{A}$ and contains the sequence $(z^{n-1})_{n\geq 1}$, it follows that T maps H^2 onto $\mathcal{A}$. □

Proof of Theorem 7.2.1 We shall prove the result for $\zeta = 1$. The general case follows by a simple change of variable $z \mapsto \zeta z$.

By definition, $\mathcal{D}_1(f) < \infty$ means that

$$\frac{1}{\pi}\int_{\mathbb{D}} |f'(z)|^2 \frac{1-|z|^2}{|z-1|^2}\, dA(z) < \infty.$$

This is equivalent to saying that $f'(z)/(z-1) \in \mathcal{A}$, where $\mathcal{A}$ is the space defined in (7.3). By Lemma 7.2.3, this happens if and only if there exists $g \in H^2$ such that $f'(z)/(z-1) = Tg(z)$. This in turn means that $f'(z) = ((z-1)g(z))'$, or equivalently, that $f(z) = a + (z-1)g(z)$ for some $a \in \mathbb{C}$. This establishes the basic equivalence.

Now assume that $f(z) = a + (z-1)g(z)$, where $g \in H^2$. Using the fact that T is an isometry from H^2 to $\mathcal{A}$, we have

$$\|g\|_{H^2}^2 = \|Tg\|_{\mathcal{A}}^2 = \left\|\frac{f'(z)}{z-1}\right\|_{\mathcal{A}}^2 = \mathcal{D}_1(f).$$

Also, by Lemma 1.5.4 we have $\lim_{|z|\to 1}(1-|z|^2)|g(z)| = 0$, and so

$$|f(z) - a|^2 = |z-1|^2|g(z)|^2 = o\Big(\frac{|z-1|^2}{1-|z|^2}\Big) \quad (|z| \to 1).$$

Therefore $f(z) \to a$ as $z \to 1$ in any approach region $|z-1| < \kappa(1-|z|^2)^{1/2}$. □

Corollary 7.2.4 *If $f \in \mathcal{D}_\mu$, then $f^*(\zeta)$ exists μ-a.e. on $\mathbb{T}$.*

Proof Let $f \in \mathcal{D}_\mu$. By Theorem 7.1.3, $\mathcal{D}_\zeta(f) < \infty$ for μ-almost every $\zeta \in \mathbb{T}$, and by Theorem 7.2.1, $f^*(\zeta)$ exists at each such ζ. □

It is now a simple matter to derive a version of Douglas' formula for $\mathcal{D}_\zeta$.

Theorem 7.2.5 (Local Douglas formula) *Let $f \in H^2$ and let $\zeta \in \mathbb{T}$. If $f^*(\zeta)$ exists, then*

$$\mathcal{D}_\zeta(f) = \frac{1}{2\pi} \int_{\mathbb{T}} \frac{|f^*(\lambda) - f^*(\zeta)|^2}{|\lambda - \zeta|^2} |d\lambda|. \tag{7.4}$$

Otherwise $\mathcal{D}_\zeta(f) = \infty$.

Of course, the original Douglas formula Theorem 1.5.1 follows upon integrating both sides of (7.4) with respect to Lebesgue measure and using (7.2).

Proof of Theorem 7.2.5 Suppose first that $\mathcal{D}_\zeta(f) < \infty$. Then $f^*(\zeta)$ exists by Theorem 7.2.1, and $\mathcal{D}_\zeta(f) = \|g\|^2_{H^2}$, where $g(z) := (f(z) - f^*(\zeta))/(z - \zeta)$. So (7.4) certainly holds in this case.

It remains to prove that, if $f^*(\zeta)$ exists and $\mathcal{D}_\zeta(f) = \infty$, then the right-hand side of (7.4) is infinite. Consider $g(z) := (f(z) - f^*(\zeta))/(z - \zeta)$. The numerator of g is in H^2, and the denominator is an outer function. If we had $g^* \in L^2(\mathbb{T})$, then by the Smirnov maximum principle (see Theorem A.3.8 in Appendix A) it would follow that $g \in H^2$, and Theorem 7.2.1 would imply that $\mathcal{D}_\zeta(f) < \infty$, contrary to hypothesis. Therefore $\|g^*\|_{L^2(\mathbb{T})} = \infty$, which amounts to saying that the right-hand side of (7.4) is infinite, as required. □

Using Theorem 7.2.1, or rather the ideas in its proof, we can also derive a formula for $\mathcal{D}_\zeta(f)$ in terms of the Taylor coefficients of f.

Theorem 7.2.6 *Let $f \in \mathrm{Hol}(\mathbb{D})$, say $f(z) = \sum_{k\geq 0} a_k z^k$, and let $\zeta \in \mathbb{T}$. Then*

$$\mathcal{D}_\zeta(f) = \sum_{n\geq 1} \frac{1}{n(n+1)} \Big|\sum_{k=1}^{n} k a_k \zeta^k\Big|^2. \tag{7.5}$$

Proof Making the change of variable $z \mapsto \zeta z$, we can reduce to the case $\zeta = 1$. So we just need to prove that

$$\mathcal{D}_1(f) = \sum_{n\geq 1} \frac{1}{n(n+1)} \Big|\sum_{k=1}^{n} k a_k\Big|^2. \tag{7.6}$$

Suppose first that $\mathcal{D}_1(f) < \infty$. By Theorem 7.2.1, $f(z) = a + (z-1)g(z)$, where $g \in H^2$ and $\mathcal{D}_1(f) = \|g\|^2_{H^2}$. We can compute $\|g\|^2_{H^2}$ via Parseval's formula, using the orthogonal basis $(p_n)_{n\geq 1}$ of H^2 furnished by Lemma 7.2.2:

$$\|g\|^2_{H^2} = \sum_{n\geq 1} \frac{|\langle g, p_n\rangle_{H^2}|^2}{\|p_n\|^2_{H^2}} = \sum_{n\geq 1} \frac{|\langle g, p_n\rangle_{H^2}|^2}{n(n+1)}.$$

Now, writing $g(z) = \sum_{k\geq 0} b_k z^k$, we have $\langle g, p_n\rangle_{H^2} = nb_n - \sum_{k=0}^{n-1} b_k$. Also, as $f(z) = a + (z-1)g(z)$, we have $a_0 = a - b_0$ and $a_k = b_{k-1} - b_k$ for all $k \geq 1$. It follows that $\langle g, p_n\rangle_{H^2} = -\sum_{k=1}^{n} ka_k$. Combining these facts, we obtain (7.6).

Now suppose that $\mathcal{D}_1(f) = \infty$. We need to show that the right-hand side of (7.6) is infinite too. Suppose, if possible, that it is finite. Then the series

$$g := -\sum_{n\geq 1}\Big(\sum_{k=1}^{n} ka_k\Big)\frac{p_n}{n(n+1)}$$

converges in H^2. Applying the operator T defined in Lemma 7.2.3 to both sides, we obtain

$$Tg(z) = -\sum_{n\geq 1}\Big(\sum_{k=1}^{n} ka_k\Big)z^{n-1},$$

the series converging in $\mathcal{A}$, and hence locally uniformly in $\mathbb{D}$. This implies that

$$((z-1)g(z))' = -(z-1)\sum_{n\geq 1}\Big(\sum_{k=1}^{n} ka_k\Big)z^{n-1} = \sum_{k=1}^{\infty} ka_k z^{k-1} = f'(z).$$

It follows that $f(z) = a + (z-1)g(z)$ for some constant a, and so $\mathcal{D}_1(f) < \infty$, contrary to hypothesis. This completes the proof. □

Exercises 7.2

1. Let $\mathcal{A}$ be defined by (7.3). Show that, if $h \in \mathcal{A}$, then

$$|h(z)|^2 \leq \frac{2\|h\|_{\mathcal{A}}^2}{(1-|z|^2)^3} \qquad (z \in \mathbb{D}).$$

   Deduce that convergence in $\mathcal{A}$ implies local uniform convergence in $\mathbb{D}$.

2. (i) Show that, if $f \in \operatorname{Hol}(\mathbb{D})$, then, for every $\zeta \in \mathbb{T}$,

$$\|f - f(0)\|_{H^2}^2 \leq 4\mathcal{D}_\zeta(f).$$

   (ii) Let $f(z) := (z-1)(z - z^2 + z^3 + \cdots + (-1)^{n+1}z^n)$. Verify that $\mathcal{D}_1(f) = n$ and that $\|f\|_{H^2}^2 = 4n-2$. Deduce that the constant 4 in part (i) is sharp.

3. Show that integrating formula (7.5) with respect to Lebesgue measure leads to (1.1).

4. Let $f \in \operatorname{Hol}(\mathbb{D})$, say $f(z) = \sum_{k\geq 0} a_k z^k$, and suppose that $\mathcal{D}_\zeta(f) < \infty$. By Theorem 7.2.1, $f(z) = f^*(\zeta) + (z-\zeta)g(z)$, where $g \in H^2$.

   (i) Writing $g(z) = \sum_{k\geq 0} b_k z^k$, show that $\sum_{k=0}^{n} a_k\zeta^k = f^*(\zeta) - b_n\zeta^{n+1}$.

   (ii) Deduce that $\sum_{k=0}^{\infty} a_k\zeta^k$ converges to $f^*(\zeta)$.

(iii) Show that

$$\mathcal{D}_\zeta(f) = \sum_{n\geq 0}\Big|\sum_{k=n+1}^{\infty} a_k\zeta^k\Big|^2. \tag{7.7}$$

(iv) We now have two apparently different formulas for $\mathcal{D}_\zeta(f)$ in terms of the Taylor coefficients of f, namely (7.5) and (7.7). Show directly that the right-hand sides of these formulas are in fact equal.

5. (Converse to previous exercise.) Let $f \in \mathrm{Hol}(\mathbb{D})$, say $f(z) = \sum_{k\geq 0} a_k z^k$, and let $\zeta \in \mathbb{T}$. Prove that, if $\sum_{k\geq 0} a_k\zeta^k$ converges and

$$\sum_{n\geq 0}\Big|\sum_{k=n+1}^{\infty} a_k\zeta^k\Big|^2 < \infty,$$

then $\mathcal{D}_\zeta(f) < \infty$.

7.3 Approximation in $\mathcal{D}_\mu$

Given $f \in \mathrm{Hol}(\mathbb{D})$ and $0 < r < 1$, let us write f_r for the *r-dilation* of f, namely $f_r(z) := f(rz)$. Each of the functions f_r is then holomorphic in a neighborhood of $\overline{\mathbb{D}}$. It is obvious from the formula (1.1) that, if $f \in \mathcal{D}$, then $\|f_r - f\|_{\mathcal{D}} \to 0$ as $r \to 1^-$. The corresponding result for $\mathcal{D}_\mu$ is also true, but trickier to prove. This is our aim in this section.

Theorem 7.3.1 *Let μ be a positive finite measure on $\mathbb{T}$, and let $f \in \mathcal{D}_\mu$. Then $\|f_r - f\|_{\mathcal{D}_\mu} \to 0$ as $r \to 1^-$.*

The key to the proof is the following estimate.

Lemma 7.3.2 *Let μ be a positive finite measure on $\mathbb{T}$, and let $f \in \mathcal{D}_\mu$. Then $\mathcal{D}_\mu(f_r) \leq \mathcal{D}_\mu(f)$ for $0 < r < 1$.*

This in turn depends on the following lemma.

Lemma 7.3.3 *Let $f \in H^2$ and let $w \in \mathbb{D}$. Then*

$$\int_{\mathbb{T}} \frac{|f^*(\lambda) - f(w)|^2}{|\lambda - w|^2}\,|d\lambda| = \int_{\mathbb{T}} \frac{|f^*(\lambda)|^2 - |f(w)|^2}{|\lambda - w|^2}\,|d\lambda|. \tag{7.8}$$

Proof The difference between the two sides is

$$\int_{\mathbb{T}} \frac{2|f(w)|^2 - 2\,\mathrm{Re}(\overline{f(w)}f^*(\lambda))}{|\lambda - w|^2}\,|d\lambda|. \tag{7.9}$$

Now $u(z) := |f(w)|^2 - \mathrm{Re}(\overline{f(w)}f(z))$ is a harmonic function on $\mathbb{D}$, which is equal to the Poisson integral of its radial boundary values u^*. In particular,

$$\frac{1}{2\pi}\int_{\mathbb{T}} \frac{1-|w|^2}{|\lambda - w|^2} u^*(\lambda)\,|d\lambda| = u(w) = 0.$$

Since $u^*(\lambda) = |f(w)|^2 - \mathrm{Re}(\overline{f(w)}f^*(\lambda))$ a.e. on $\mathbb{T}$, it follows that the integral (7.9) vanishes, as desired. □

Proof of Lemma 7.3.2 It suffices to show that $\mathcal{D}_\zeta(f_r) \le \mathcal{D}_\zeta(f)$ for all $\zeta \in \mathbb{T}$. The result then follows by integrating both sides with respect to $d\mu(\zeta)$.

Fix $\zeta \in \mathbb{T}$. We may suppose that $\mathcal{D}_\zeta(f) < \infty$. Then, by Lemma 7.2.3, we have $f(z) = a + (z-\zeta)g(z)$, where $g \in H^2$ and $\|g\|_{H^2}^2 = \mathcal{D}_\zeta(f)$. Let $r < 1$. A simple computation shows that $f_r(z) = b + (z-\zeta)h_r(z)$, where $b := f(r\zeta)$ and

$$h(z) := r\frac{(z-\zeta)g(z) - (r\zeta - \zeta)g(r\zeta)}{z - r\zeta}.$$

By Lemma 7.2.3 again, it follows that $\mathcal{D}_\zeta(f_r) = \|h_r\|_{H^2}^2 \le \|h\|_{H^2}^2$. Now

$$\begin{aligned}
\|h\|_{H^2}^2 &= \frac{1}{2\pi}\int_{\mathbb{T}} r^2 \left|\frac{(\lambda-\zeta)g^*(\lambda) - (r\zeta-\zeta)g(r\zeta)}{\lambda - r\zeta}\right|^2 |d\lambda| \\
&\le \frac{1}{2\pi}\int_{\mathbb{T}} r^2 \left|\frac{(\lambda-\zeta)g^*(\lambda)}{\lambda - r\zeta}\right|^2 |d\lambda| \\
&\le \|g\|_{H^2}^2,
\end{aligned}$$

where the first inequality comes from Lemma 7.3.3, and the second one holds because, using Lemma 1.6.7, we have

$$\frac{|r(\lambda-\zeta)|}{|\lambda - r\zeta|} \le \frac{2r}{1+r} \le 1 \qquad (\lambda \in \mathbb{T}).$$

Putting all these estimates together, we obtain $\mathcal{D}_\zeta(f_r) \le \mathcal{D}_\zeta(f)$, as desired. □

Proof of Theorem 7.3.1 Using the parallelogram identity, we have

$$\mathcal{D}_\mu(f_r - f) + \mathcal{D}_\mu(f_r + f) = 2\mathcal{D}_\mu(f) + 2\mathcal{D}_\mu(f_r) \qquad (0 < r < 1).$$

By Lemma 7.3.2, $\mathcal{D}_\mu(f_r) \le \mathcal{D}_\mu(f)$ for all $r \in (0,1)$. Also, by Fatou's lemma, $\liminf_{r\to 1} \mathcal{D}_\mu(f_r + f) \ge \mathcal{D}_\mu(2f) = 4\mathcal{D}_\mu(f)$. Hence $\limsup_{r\to 1} \mathcal{D}_\mu(f_r - f) \le 0$. Therefore, finally,

$$\lim_{r\to 1^-} \|f_r - f\|_{\mathcal{D}_\mu}^2 = \lim_{r\to 1^-}\left(\|f_r - f\|_{H^2}^2 + \mathcal{D}_\mu(f_r - f)\right) = 0 + 0 = 0. \qquad \square$$

We conclude with a simple but important application of this result.

Corollary 7.3.4 *The polynomials are dense in $\mathcal{D}_\mu$.*

Proof Let $f \in \mathcal{D}_\mu$ and let $\epsilon > 0$. By Theorem 7.3.1, there exists $r \in (0,1)$ such that $\|f_r - f\|_{\mathcal{D}_\mu} < \epsilon/2$. Fix this r, and let $s_n f_r$ be the n-th partial sum of the Taylor expansion of f_r around zero. Since f_r is holomorphic on a neighborhood of $\overline{\mathbb{D}}$, the sequences $(s_n f_r)_{n\ge0}$ and $(s_n f_r)'_{n\ge0}$ converge uniformly on $\overline{\mathbb{D}}$ to f_r and f_r' respectively, and it follows that $\|s_n f_r - f_r\|_{\mathcal{D}_\mu} < \epsilon/2$ if n is large enough. Then $s_n f_r$ is a polynomial and $\|s_n f_r - f\|_{\mathcal{D}_\mu} < \epsilon$. □

It is not in general true, however, that the Taylor series of $f \in \mathcal{D}_\mu$ converges to f in the norm of $\mathcal{D}_\mu$. See Exercise 7.3.2 below.

Exercises 7.3

1. Let $f \in \mathrm{Hol}(\mathbb{D})$ and let $\zeta \in \mathbb{T}$. Show that $\mathcal{D}_\zeta(f_r)$ is an increasing function of $r \in (0,1)$, and that $\mathcal{D}_\zeta(f_r) \to \mathcal{D}_\zeta(f)$ as $r \to 1^-$.
2. Given $f \in \mathrm{Hol}(\mathbb{D})$, denote by $s_n f$ the n-th partial sum of the Taylor expansion of f about zero.
   (i) Show that, if $f(z) := z^n - z^{n+1}$, then $\|f\|^2_{\mathcal{D}_1} = 3$ and $\|s_n f\|^2_{\mathcal{D}_1} = n+1$.
   (ii) Deduce that there exists $f \in \mathcal{D}_1$ such that $\sup_{n\ge0} \|s_n f\|_{\mathcal{D}_1} = \infty$. In particular, $s_n f \not\to f$ in $\mathcal{D}_1$. [Hint: Banach–Steinhaus theorem.]

7.4 Outer functions

Recall that $f \in \mathrm{Hol}(\mathbb{D})$ is an *outer function* if it is of the form

$$f(z) = \exp\Big(\frac{1}{2\pi}\int_{\mathbb{T}} \frac{\zeta + z}{\zeta - z} \log\phi(\zeta)\,|d\zeta|\Big) \qquad (z \in \mathbb{D}),$$

where $\phi : \mathbb{T} \to \mathbb{R}^+$ is a function such that $\log\phi \in L^1(\mathbb{T})$. In this case the radial limit $f^*(\zeta) := \lim_{r\to1^-} f(r\zeta)$ exists for a.e. $\zeta \in \mathbb{T}$, and $|f^*| = \phi$ a.e. Also $f \in H^2$ if and only if $\phi \in L^2(\mathbb{T})$. For more information about outer functions, see Appendix A.2.

An outer function f is uniquely determined by $|f^*|$, and this is frequently how it is specified in practice. It is therefore desirable to be able to express $\mathcal{D}(f)$, or more generally $\mathcal{D}_\mu(f)$, purely in terms of $|f^*|$ (as opposed to f^*). Notice that Douglas' formula does not achieve this, because the difference $|f^*(\lambda) - f^*(\zeta)|$ depends not only on the moduli of $f^*(\lambda)$ and $f^*(\zeta)$, but also on their arguments, arguments which may be difficult to determine in practice.

A formula for $\mathcal{D}(f)$ of the type that we seek was discovered by Carleson, and subsequently extended to the case of $\mathcal{D}_\zeta(f)$ by Richter and Sundberg. Here are their results.

Theorem 7.4.1 (Carleson's formula) *Let $f \in H^2$ be an outer function. Then*

$$\mathcal{D}(f) = \frac{1}{4\pi^2}\int_{\mathbb{T}}\int_{\mathbb{T}} \frac{(|f^*(\lambda)|^2 - |f^*(\zeta)|^2)(\log|f^*(\lambda)| - \log|f^*(\zeta)|)}{|\lambda-\zeta|^2}\,|d\lambda|\,|d\zeta|.$$

Theorem 7.4.2 (Richter–Sundberg formula) *Let $f \in H^2$ be an outer function, and let $\zeta \in \mathbb{T}$ be a point such that $f^*(\zeta)$ exists. Then*

$$\mathcal{D}_\zeta(f) = \frac{1}{2\pi}\int_{\mathbb{T}} \frac{|f^*(\lambda)|^2 - |f^*(\zeta)|^2 - 2|f^*(\zeta)|^2\log|f^*(\lambda)/f^*(\zeta)|}{|\lambda-\zeta|^2}\,|d\lambda|. \tag{7.10}$$

Remarks (i) For certain classes of functions, Carleson's formula can be used to derive estimates for the Dirichlet integral which, though not exact formulas, are of a much simpler form. A useful example of this kind, for so-called distance functions, will be given in Theorem 9.4.3.

(ii) Formula (7.10) contains an ambiguity if $f^*(\zeta) = 0$. In this case, the product $|f^*(\zeta)|^2\log|f^*(\lambda)/f^*(\zeta)|$ should be interpreted as being zero.

(iii) The integrands in both formulas are non-negative, so the integrals are well defined, possibly infinite.

(iv) If $f \notin H^2$ or if the radial limit $f^*(\zeta)$ does not exist, then $\mathcal{D}_\zeta(f) = \infty$, as we already saw in Theorem 7.2.1.

(v) Of course, given a finite measure μ on $\mathbb{T}$, we can get a formula for $\mathcal{D}_\mu(f)$ as usual, by integrating both sides of (7.10) with respect to $d\mu(\zeta)$. The case when μ is Lebesgue measure leads to the following proof of Theorem 7.4.1.

Proof of Theorem 7.4.1 Since $f \in H^2$, it follows that $f^*(\zeta)$ exists and is non-zero for a.e. $\zeta \in \mathbb{T}$. Thus (7.10) holds for a.e. $\zeta \in \mathbb{T}$. Integrating both sides with respect to normalized Lebesgue measure, we find that

$$\mathcal{D}(f) = \frac{1}{4\pi^2}\int_{\mathbb{T}}\int_{\mathbb{T}} \frac{|f^*(\lambda)|^2 - |f^*(\zeta)|^2 - 2|f^*(\zeta)|^2\log|f^*(\lambda)/f^*(\zeta)|}{|\lambda-\zeta|^2}\,|d\lambda|\,|d\zeta|.$$

If we exchange the roles of λ and ζ, then we also have

$$\mathcal{D}(f) = \frac{1}{4\pi^2}\int_{\mathbb{T}}\int_{\mathbb{T}} \frac{|f^*(\zeta)|^2 - |f^*(\lambda)|^2 - 2|f^*(\lambda)|^2\log|f^*(\zeta)/f^*(\lambda)|}{|\lambda-\zeta|^2}\,|d\lambda|\,|d\zeta|.$$

Taking the average of these two expressions for $\mathcal{D}(f)$, we obtain the formula in the statement of the theorem. □

It remains to prove Theorem 7.4.2. This will occupy us for much of the rest of the section. The main idea is to re-express the local Douglas formula (7.4) purely in terms of $|f^*|$. Our starting point in this endeavor is Lemma 7.3.3. Recall that it says that, if $f \in H^2$ and $w \in \mathbb{D}$, then

$$\int_{\mathbb{T}} \frac{|f^*(\lambda) - f(w)|^2}{|\lambda - w|^2}\,|d\lambda| = \int_{\mathbb{T}} \frac{|f^*(\lambda)|^2 - |f(w)|^2}{|\lambda - w|^2}\,|d\lambda|. \tag{7.11}$$

The plan is to take $w = r\zeta$ in both sides of (7.11) and let $r \to 1$. We begin with the left-hand side.

Lemma 7.4.3 *Let $f \in H^2$ and let $\zeta \in \mathbb{T}$. Then*

$$\lim_{r\to 1} \int_{\mathbb{T}} \frac{|f^*(\lambda) - f(r\zeta)|^2}{|\lambda - r\zeta|^2} |d\lambda| = 2\pi \mathcal{D}_\zeta(f).$$

Proof We first show that

$$\liminf_{r\to 1} \int_{\mathbb{T}} \frac{|f^*(\lambda) - f(r\zeta)|^2}{|\lambda - r\zeta|^2} |d\lambda| \geq 2\pi \mathcal{D}_\zeta(f). \tag{7.12}$$

We may assume that the left-hand side is finite. Pick a sequence $r_n \to 1$ such that the family of functions $g_n(z) := (f(z) - f(r_n\zeta))/(z - r_n\zeta)$ satisfies

$$\lim_{n\to\infty} \|g_n\|_{H^2}^2 = \liminf_{r\to 1} \frac{1}{2\pi} \int_{\mathbb{T}} \frac{|f^*(\lambda) - f(r\zeta)|^2}{|\lambda - r\zeta|^2} |d\lambda|.$$

A subsequence (g_{n_k}) then converges weakly in H^2 to a limit g. In particular, $g_{n_k}(0) \to g(0)$, so the corresponding subsequence of scalars $f(r_{n_k}\zeta)$ converges to a limit a, and we have $f(z) = a + (z - \zeta)g(z)$. Using Theorem 7.2.1, we deduce that

$$\mathcal{D}_\zeta(f) = \|g\|_{H^2}^2 \leq \lim_{k\to\infty} \|g_{n_k}\|_{H^2}^2,$$

which gives (7.12).

Now we show that

$$\limsup_{r\to 1} \int_{\mathbb{T}} \frac{|f^*(\lambda) - f(r\zeta)|^2}{|\lambda - r\zeta|^2} |d\lambda| \leq 2\pi \mathcal{D}_\zeta(f). \tag{7.13}$$

We may assume that the right-hand side is finite. Then, in particular, $f^*(\zeta)$ exists. Lemma 7.3.3, applied with f replaced by $f - f^*(\zeta)$, gives

$$\int_{\mathbb{T}} \frac{|f^*(\lambda) - f(w)|^2}{|\lambda - w|^2} |d\lambda| \leq \int_{\mathbb{T}} \frac{|f^*(\lambda) - f^*(\zeta)|^2}{|\lambda - w|^2} |d\lambda| \qquad (w \in \mathbb{D}).$$

It follows that

$$\begin{aligned}\limsup_{r\to 1} \int_{\mathbb{T}} \frac{|f^*(\lambda) - f(r\zeta)|^2}{|\lambda - r\zeta|^2} |d\lambda| &\leq \limsup_{r\to 1} \int_{\mathbb{T}} \frac{|f^*(\lambda) - f^*(\zeta)|^2}{|\lambda - r\zeta|^2} |d\lambda| \\ &\leq \int_{\mathbb{T}} \frac{|f^*(\lambda) - f^*(\zeta)|^2}{|\lambda - \zeta|^2} |d\lambda|,\end{aligned}$$

where the second inequality comes from the dominated convergence theorem (justified using the inequality of Lemma 1.6.7). This establishes (7.13). □

Now we turn our attention to the right-hand side of (7.11). It is here that we need to assume that f is outer.

Lemma 7.4.4 *Let $f \in H^2$ be an outer function, and let $\zeta \in \mathbb{T}$ be a point such that* $\lim_{r\to 1} |f(r\zeta)| = 1$. *Then*

$$\lim_{r\to 1} \int_{\mathbb{T}} \frac{|f^*(\lambda)|^2 - |f(r\zeta)|^2}{|\lambda - r\zeta|^2}\,|d\lambda| = \int_{\mathbb{T}} \frac{|f^*(\lambda)|^2 - 1 - 2\log|f^*(\lambda)|}{|\lambda - \zeta|^2}\,|d\lambda|.$$

Proof Since f is an outer function, $\log|f|$ is the Poisson integral of $\log|f^*|$. It follows that

$$\int_{\mathbb{T}} \frac{\log|f^*(\lambda)/f(r\zeta)|}{|\lambda - r\zeta|^2}\,|d\lambda| = 0 \qquad (0 < r < 1). \tag{7.14}$$

Hence,

$$\begin{aligned}
&\int_{\mathbb{T}} \frac{|f^*(\lambda)|^2 - |f(r\zeta)|^2}{|\lambda - r\zeta|^2}\,|d\lambda| \\
&= \int_{\mathbb{T}} \frac{|f^*(\lambda)|^2 - |f(r\zeta)|^2 - 2|f(r\zeta)|^2\log|f^*(\lambda)/f(r\zeta)|}{|\lambda - r\zeta|^2}\,|d\lambda| \qquad (0 < r < 1).
\end{aligned}$$

Now $|f(r\zeta)| \to 1$ as $r \to 1$. Also, the integrands on the right-hand side are all positive, because $x - 1 - \log x \geq 0$ for all $x > 0$. Therefore we may apply Fatou's lemma, to deduce that

$$\liminf_{r\to 1} \int_{\mathbb{T}} \frac{|f^*(\lambda)|^2 - |f(r\zeta)|^2}{|\lambda - r\zeta|^2}\,|d\lambda| \geq \int_{\mathbb{T}} \frac{|f^*(\lambda)|^2 - 1 - 2\log|f^*(\lambda)|}{|\lambda - \zeta|^2}\,|d\lambda|.$$

In the other direction, we begin by observing that, for all $a, b > 0$,

$$\begin{aligned}
a^2 - 1 - 2\log a &= (a^2 - b^2) + (b^2 - 1 - 2\log b) - 2\log(a/b) \\
&\geq a^2 - b^2 - 2\log(a/b),
\end{aligned}$$

and thus

$$|f^*(\lambda)|^2 - 1 - 2\log|f^*(\lambda)| \geq |f^*(\lambda)|^2 - |f(r\zeta)|^2 - 2\log|f^*(\lambda)/f(r\zeta)|).$$

The last term on the right-hand side satisfies (7.14). Therefore we have

$$\int_{\mathbb{T}} \frac{|f^*(\lambda)|^2 - 1 - 2\log|f^*(\lambda)|}{|\lambda - r\zeta|^2}\,|d\lambda| \geq \int_{\mathbb{T}} \frac{|f^*(\lambda)|^2 - |f(r\zeta)|^2}{|\lambda - r\zeta|^2}\,|d\lambda| \quad (0 < r < 1).$$

It follows that

$$\begin{aligned}
\limsup_{r\to 1} \int_{\mathbb{T}} \frac{|f^*(\lambda)|^2 - |f(r\zeta)|^2}{|\lambda - r\zeta|^2}\,|d\lambda| &\leq \limsup_{r\to 1} \int_{\mathbb{T}} \frac{|f^*(\lambda)|^2 - 1 - 2\log|f^*(\lambda)|}{|\lambda - r\zeta|^2}\,|d\lambda| \\
&\leq \int_{\mathbb{T}} \frac{|f^*(\lambda)|^2 - 1 - 2\log|f^*(\lambda)|}{|\lambda - \zeta|^2}\,|d\lambda|.
\end{aligned}$$

Here the last inequality follows from the dominated convergence theorem if the final integral is finite (justified again using Lemma 1.6.7), and is obvious anyway if the integral is infinite. □

Now we can prove the Richter–Sundberg formula.

Proof of Theorem 7.4.2 Suppose first that $f^*(\zeta) \neq 0$. Replacing f by $f/f^*(\zeta)$, we can suppose that $f^*(\zeta) = 1$. Set $w = r\zeta$ in (7.11), and let $r \to 1$ using Lemmas 7.4.3 and 7.4.4. This yields

$$2\pi\mathcal{D}_\zeta(f) = \int_{\mathbb{T}} \frac{|f^*(\lambda)|^2 - 1 - 2\log|f^*(\lambda)|}{|\lambda - \zeta|^2}\,|d\lambda|,$$

giving (7.10) in this case.

If $f^*(\zeta) = 0$, then the right-hand side of (7.10) is the same as the right-hand side of (7.4), so the result follows directly from Theorem 7.2.5 in this case. □

The final result of this section complements the Richter–Sundberg formula.

Theorem 7.4.5 *Let f be an outer function (not necessarily in H^2). Let $\zeta \in \mathbb{T}$, let $a \geq 0$, and suppose that*

$$\int_{\mathbb{T}} \frac{|f^*(\lambda)|^2 - a^2 - 2a^2\log|f^*(\lambda)/a|}{|\lambda - \zeta|^2}\,|d\lambda| < \infty.$$

Then $\mathcal{D}_\zeta(f) < \infty$, and $f^*(\zeta)$ exists and satisfies $|f^*(\zeta)| = a$.

Remark The term $2a^2\log|f^*(\lambda)/a|$ should be interpreted as being zero if $a = 0$.

Proof The case when $a > 0$ can be reduced to $a = 1$ by considering f/a. There are thus two cases to consider, namely $a = 0$ and $a = 1$.

If $a = 0$, then the hypothesis becomes

$$\int_{\mathbb{T}} \frac{|f^*(\lambda)|^2}{|\lambda - \zeta|^2}\,|d\lambda| < \infty.$$

Thus $f(z) = (z - \zeta)g(z)$ where $g \in H^2$, and so by Theorem 7.2.1 we have $\mathcal{D}_\zeta(f) < \infty$ and $f^*(\zeta) = 0$, proving the result in this case.

For the rest of the proof, we suppose that $a = 1$. The hypothesis becomes

$$\int_{\mathbb{T}} \frac{|f^*(\lambda)|^2 - 1 - 2\log|f^*(\lambda)|}{|\lambda - \zeta|^2}\,|d\lambda| < \infty. \tag{7.15}$$

The first step is to show that (7.15) implies that $f \in H^2$. For this, we use the inequality $x^2 - 1 - 2\log x \geq (x-1)^2$, which, together with (7.15), gives

$$\int_{\mathbb{T}} \frac{(|f^*(\lambda)| - 1)^2}{|\lambda - \zeta|^2}\,|d\lambda| < \infty.$$

This easily implies that $f^* \in L^2(\mathbb{T})$. Hence, by Smirnov's maximum principle (see Theorem A.3.8 in Appendix A), we do indeed have $f \in H^2$.

Next, we show that $\lim_{r\to 1}|f(r\zeta)| = 1$. For this, let us recall that, since f is outer, $\log|f|$ is the Poisson integral of $\log|f^*|$. Hence, if $\phi : \mathbb{R} \to \mathbb{R}^+$ is any positive convex function, then, by Jensen's inequality, we have

$$\phi(\log|f(r\zeta)|) \leq \frac{1}{2\pi}\int_{\mathbb{T}} \phi(\log|f^*(\lambda)|)\frac{1-r^2}{|\lambda - r\zeta|^2}\,|d\lambda|$$
$$\leq \frac{1-r^2}{2\pi}\int_{\mathbb{T}} \phi(\log|f^*(\lambda)|)\frac{4}{|\lambda-\zeta|^2}\,|d\lambda|.$$

Let us apply this inequality with $\phi(x) := e^{2x} - 1 - 2x$, which is indeed both positive and convex. The last integral is then finite by (7.15). It follows that $\lim_{r\to 1}\phi(\log|f(r\zeta)|) = 0$, which in turn forces $\lim_{r\to 1}|f(r\zeta)| = 1$, as claimed.

To conclude, we use successively Lemmas 7.4.3, 7.3.3 and 7.4.4 to obtain

$$2\pi\mathcal{D}_\zeta(f) = \lim_{r\to 1}\int_{\mathbb{T}} \frac{|f^*(\lambda) - f(r\zeta)|^2}{|\lambda - r\zeta|^2}\,|d\lambda|$$
$$= \lim_{r\to 1}\int_{\mathbb{T}} \frac{|f^*(\lambda)|^2 - |f(r\zeta)|^2}{|\lambda - r\zeta|^2}\,|d\lambda|$$
$$= \int_{\mathbb{T}} \frac{|f^*(\lambda)|^2 - 1 - 2\log|f^*(\lambda)|}{|\lambda-\zeta|^2}\,|d\lambda|.$$

In particular $\mathcal{D}_\zeta(f) < \infty$. By Theorem 7.2.1 $f^*(\zeta)$ exists and, from what we have already proved, $|f^*(\zeta)| = \lim_{r\to 1}|f(r\zeta)| = 1$. The proof is complete. □

Exercises 7.4

1. Let $f \in \mathrm{Hol}(\mathbb{D})$ be an outer function. Show that, if $f \notin H^2$, then Carleson's formula (Theorem 7.4.1) continues to hold in the sense that

$$\int_{\mathbb{T}}\int_{\mathbb{T}} \frac{(|f^*(\lambda)|^2 - |f^*(\zeta)|^2)(\log|f^*(\lambda)| - \log|f^*(\zeta)|)}{|\lambda-\zeta|^2}\,|d\lambda|\,|d\zeta| = \infty.$$

   [Hint: Use Theorem 7.4.5.]

7.5 Lattice operations in $\mathcal{D}_\mu$

Let $\vee, \wedge$ denote the usual lattice operations in $\mathbb{R}^+$, namely $x \vee y := \max\{x, y\}$ and $x \wedge y := \min\{x, y\}$. We extend these operations to outer functions as follows.

Definition 7.5.1 Given outer functions f, g, we define $f \vee g$ and $f \wedge g$ to be the unique outer functions satisfying the following relations a.e. on $\mathbb{T}$:

$$|(f \vee g)^*(\zeta)| = |f^*(\zeta)| \vee |g^*(\zeta)|,$$
$$|(f \wedge g)^*(\zeta)| = |f^*(\zeta)| \wedge |g^*(\zeta)|.$$

We shall show that the $\mathcal{D}_\mu$-spaces are stable under these operations. In particular, this is true of $\mathcal{D}$.

Theorem 7.5.2 *Let μ be a finite measure on $\mathbb{T}$ and let f, g be outer functions in $\mathcal{D}_\mu$. Then $f \vee g$ and $f \wedge g$ belong to $\mathcal{D}_\mu$, and*

$$\mathcal{D}_\mu(f \vee g) \le \mathcal{D}_\mu(f) + \mathcal{D}_\mu(g),$$
$$\mathcal{D}_\mu(f \wedge g) \le \mathcal{D}_\mu(f) + \mathcal{D}_\mu(g).$$

The proof depends on the results of the previous section and the following elementary lemma.

Lemma 7.5.3 *Define $F : (0, \infty) \times [0, \infty) \to \mathbb{R}^+$ by*

$$F(x, y) := \begin{cases} x - y - y\log(x/y), & x > 0,\ y > 0, \\ x, & x > 0,\ y = 0. \end{cases}$$

Then, for all $x_1, x_2 \in (0, \infty)$ and $y_1, y_2 \in [0, \infty)$, we have

$$F(x_1 \vee x_2,\ y_1 \vee y_2) \le F(x_1, y_1) + F(x_2, y_2),$$
$$F(x_1 \wedge x_2,\ y_1 \wedge y_2) \le F(x_1, y_1) + F(x_2, y_2).$$

Proof We first establish a monotonicity property of F. A simple computation gives $\partial F/\partial x = 1 - y/x$ and $\partial F/\partial y = \log(y/x)$. It follows that the line $y = x$ is the valley of the graph of F in the following sense:

- $x \mapsto F(x, y)$ is decreasing for $x \le y$ and increasing for $x \ge y$,
- $y \mapsto F(x, y)$ is decreasing for $y \le x$ and increasing for $y \ge x$.

For the main part of the proof, we can suppose without loss of generality that $x_1 \le x_2$. If $y_1 \le y_2$, then the result is obvious. Suppose that $y_1 \ge y_2$. Using the monotonicity property of F above, we then have

$$F(x_1 \vee x_2,\ y_1 \vee y_2) = F(x_2, y_1) \le \begin{cases} F(x_1, y_1), & \text{if } x_2 \le y_1, \\ F(x_2, y_2), & \text{if } x_2 \ge y_1, \end{cases}$$

and

$$F(x_1 \wedge x_2,\ y_1 \wedge y_2) = F(x_1, y_2) \le \begin{cases} F(x_1, y_1), & \text{if } x_1 \le y_2, \\ F(x_2, y_2), & \text{if } x_1 \ge y_2. \end{cases}$$

As F is non-negative, the terms on the right-hand side are all bounded above by $F(x_1, y_1) + F(x_2, y_2)$. □

Proof of Theorem 7.5.2 Since $f, g \in \mathcal{D}_\mu$, we have $\mathcal{D}_\zeta(f) + \mathcal{D}_\zeta(g) < \infty$ for μ-almost every $\zeta \in \mathbb{T}$. Fix such a ζ. In particular $f^*(\zeta)$ and $g^*(\zeta)$ both exist. Define $a := |f^*(\zeta)| \vee |g^*(\zeta)|$. Then, using Lemma 7.5.3 and Theorem 7.4.2, we have

$$\begin{aligned}
&\int_{\mathbb{T}} \frac{|(f \vee g)^*(\lambda)|^2 - a^2 - 2a^2 \log|(f \vee g)^*(\lambda)/a|}{|\lambda - \zeta|^2} \,|d\lambda| \\
&= \int_{\mathbb{T}} \frac{F(|f^*(\lambda)|^2 \vee |g^*(\lambda)|^2, |f^*(\zeta)|^2 \vee |g^*(\zeta)|^2)}{|\lambda - \zeta|^2} \,|d\lambda| \\
&\le \int_{\mathbb{T}} \frac{F(|f^*(\lambda)|^2, |f^*(\zeta)|^2)}{|\lambda - \zeta|^2} \,|d\lambda| + \int_{\mathbb{T}} \frac{F(|g^*(\lambda)|^2, |g^*(\zeta)|^2)}{|\lambda - \zeta|^2} \,|d\lambda| \\
&= 2\pi \mathcal{D}_\zeta(f) + 2\pi \mathcal{D}_\zeta(g) < \infty.
\end{aligned}$$

By Theorem 7.4.5, it follows that $(f \vee g)^*(\zeta)$ exists and satisfies $|(f \vee g)^*(\zeta)| = a$. Substituting this information into (7.10) and repeating the calculation above, we obtain that $\mathcal{D}_\zeta(f \vee g) \le \mathcal{D}_\zeta(f) + \mathcal{D}_\zeta(g)$. Integrating with respect to $d\mu(\zeta)$ gives $\mathcal{D}_\mu(f \vee g) \le \mathcal{D}_\mu(f) + \mathcal{D}_\mu(g)$.

An exactly analogous argument works for $f \wedge g$. □

In general $f \in \mathcal{D}_\mu$ does not imply $f^2 \in \mathcal{D}_\mu$ (see Exercise 7.1.4). However, we do have the following restricted version.

Theorem 7.5.4 *Let μ be a finite measure on $\mathbb{T}$, and let $f \in \mathcal{D}_\mu$ be outer. Then $f \wedge f^2 \in \mathcal{D}_\mu$ and*

$$\mathcal{D}_\mu(f \wedge f^2) \le 4\mathcal{D}_\mu(f).$$

For the proof, we need another elementary lemma, similar to Lemma 7.5.3.

Lemma 7.5.5 *Define $F : (0, \infty) \times [0, \infty) \to \mathbb{R}^+$ as in Lemma 7.5.3. Then, for all $x \in (0, \infty)$ and $y \in [0, \infty)$, we have*

$$F(x \wedge x^2, y \wedge y^2) \le 4F(x, y).$$

Proof Set $G(x, y) := 4F(x, y) - F(x \wedge x^2, y \wedge y^2)$. We show that $G(x, y) \ge 0$. There are four cases to consider.

Case 1: $x \ge 1$ and $y \ge 1$. In this case, we have

$$G(x, y) = 4F(x, y) - F(x, y) = 3F(x, y) \ge 0.$$

Case 2: $x \ge 1$ and $y < 1$. Writing G_y for $\partial G/\partial y$, for all $t < 1$ we have $G_y(x, t) = -4(1 - t)\log(x/t) - 2t \log x \le 0$. It follows that

$$G(x, y) \ge G(x, 1) \ge 0,$$

where the last inequality is from Case 1.

Case 3: $x < 1$ and $y \leq 1$. For all $t < 1$ we have $G_y(x,t) = 4(1-t)\log(t/x)$, which has the same sign as $t - x$. It follows that

$$G(x,y) \geq G(x,x) = 0,$$

where the last equality holds since $F(t,t) = 0$ for all t.

Case 4: $x < 1$ and $y > 1$. For $t > 1$, we have $G_y(x,t) = 3\log t - 2\log x \geq 0$. It follows that

$$G(x,y) \geq G(x,1) \geq 0,$$

where the last inequality is from Case 3. □

Proof of Theorem 7.5.4 This is just like the proof of Theorem 7.5.2, using Lemma 7.5.5 instead of Lemma 7.5.3. □

Exercises 7.5

1. Let $f \in \mathcal{D}_\mu$ be an outer function. Show that $f = f_1/f_2$, where f_1, f_2 are bounded functions in $\mathcal{D}_\mu$. [Hint: Take $f_1 := f \wedge 1$ and $f_2 := 1/(f \vee 1)$.]

7.6 Inner functions

Theorems 7.4.1 and 7.4.2 are valid only for outer functions: just consider what happens when $f(z) = z^n$, for example. In this section we shall derive extensions of these formulas, also due to Carleson and to Richter and Sundberg, which are valid for all functions in H^2. For these, we need to take into account inner factors.

Recall that a function θ is *inner* if it is a bounded holomorphic function on $\mathbb{D}$ such that $|\theta^*| = 1$ a.e. on $\mathbb{T}$. Every $f \in H^2 \setminus \{0\}$ has a unique factorization $f = f_i f_o$, where $f_i, f_o \in H^2$, f_i is inner and f_o is outer. For more about inner functions, see Appendix A.2.

Our starting point is a result which permits us to factor out inner functions in local Dirichlet integrals.

Theorem 7.6.1 *Let $f \in H^2$, let θ be an inner function, and let $\zeta \in \mathbb{T}$. Then*

$$\mathcal{D}_\zeta(f\theta) = \mathcal{D}_\zeta(f) + |f^*(\zeta)|^2 \mathcal{D}_\zeta(\theta). \tag{7.16}$$

Remarks (i) If the radial limit $f^*(\zeta)$ does not exist, then $\mathcal{D}_\zeta(f) = \infty$, and the right-hand side of (7.16) should be interpreted as being ∞.

(ii) If the radial limit $f^*(\zeta) = 0$, then the product $|f^*(\zeta)|^2\mathcal{D}_\zeta(\theta)$ should be interpreted as being 0, even if $\mathcal{D}_\zeta(\theta) = \infty$.

Similar remarks will apply to all the subsequent applications of this result.

Proof of Theorem 7.6.1 Using Lemma 7.3.3, for all $w \in \mathbb{D}$ we have

$$\begin{aligned}
&\int_{\mathbb{T}} \frac{|f^*(\lambda)\theta^*(\lambda) - f(w)\theta(w)|^2}{|\lambda - w|^2}\,|d\lambda| \\
&= \int_{\mathbb{T}} \frac{|f^*(\lambda)\theta^*(\lambda)|^2 - |f(w)\theta(w)|^2}{|\lambda - w|^2}\,|d\lambda| \\
&= \int_{\mathbb{T}} \frac{|f^*(\lambda)\theta^*(\lambda)|^2 - |f(w)|^2}{|\lambda - w|^2}\,|d\lambda| + \int_{\mathbb{T}} \frac{|f(w)|^2 - |f(w)\theta(w)|^2}{|\lambda - w|^2}\,|d\lambda| \\
&= \int_{\mathbb{T}} \frac{|f^*(\lambda)|^2 - |f(w)|^2}{|\lambda - w|^2}\,|d\lambda| + |f(w)|^2 \int_{\mathbb{T}} \frac{|\theta^*(\lambda)|^2 - |\theta(w)|^2}{|\lambda - w|^2}\,|d\lambda| \\
&= \int_{\mathbb{T}} \frac{|f^*(\lambda) - f(w)|^2}{|\lambda - w|^2}\,|d\lambda| + |f(w)|^2 \int_{\mathbb{T}} \frac{|\theta^*(\lambda) - \theta(w)|^2}{|\lambda - w|^2}\,|d\lambda|.
\end{aligned}$$

Put $w = r\zeta$, let $r \to 1$, and apply Lemma 7.4.3. If $f^*(\zeta)$ exists and is non-zero, then we straightaway obtain the result. If $f^*(\zeta) = 0$, then $\mathcal{D}_\zeta(f\theta) = \mathcal{D}_\zeta(f)$ directly from Theorem 7.2.5. Finally, if $f^*(\zeta)$ does not exist, then $\mathcal{D}_\zeta(f) = \infty$, and since the chain of equalities above implies that $\mathcal{D}_\zeta(f\theta) \geq \mathcal{D}_\zeta(f)$, we have $\mathcal{D}_\zeta(f\theta) = \infty$ as well. □

Corollary 7.6.2 *Let $f \in H^2$ and let θ be an inner function. If $\theta f \in \mathcal{D}_\mu$, then $f \in \mathcal{D}_\mu$, and $\mathcal{D}_\mu(f) \leq \mathcal{D}_\mu(\theta f)$.*

Proof By the theorem, $\mathcal{D}_\zeta(f) \leq \mathcal{D}_\zeta(f\theta)$ for all $\zeta \in \mathbb{T}$. Integrate both sides with respect to μ to get the result. □

The following special case is particularly worthy of note.

Corollary 7.6.3 *Let μ be a finite measure on $\mathbb{T}$. If $f \in \mathcal{D}_\mu$ and if $f = f_i f_o$ is its inner-outer factorization, then $f_o \in \mathcal{D}_\mu$ and $\mathcal{D}_\mu(f_o) \leq \mathcal{D}_\mu(f)$.*

Proof Apply the Corollary 7.6.2 with f, θ replaced by f_o, f_i respectively. □

Here is one further consequence of Theorem 7.6.1, which will prove useful later on. We recall that the notation $f \wedge g$ was defined in the previous section.

Corollary 7.6.4 *Let μ be a finite measure on $\mathbb{T}$, let $f \in H^2$ be an outer function and let θ be an inner function. If $\theta f \in \mathcal{D}_\mu$, then $\theta(f \wedge 1) \in \mathcal{D}_\mu$ and $\mathcal{D}_\mu(\theta(f \wedge 1)) \leq \mathcal{D}_\mu(\theta f)$.*

Proof By Theorems 7.6.1 and 7.5.2, for μ-almost every $\zeta \in \mathbb{T}$, we have

$$\begin{aligned}
\mathcal{D}_\zeta(\theta(f \wedge 1)) &= \mathcal{D}_\zeta(\theta)(|f^*(\zeta)| \wedge 1)^2 + \mathcal{D}_\zeta(f \wedge 1) \\
&\leq \mathcal{D}_\zeta(\theta)|f^*(\zeta)|^2 + \mathcal{D}_\zeta(f) \\
&= \mathcal{D}_\zeta(\theta f).
\end{aligned}$$

Integrating with respect to μ gives the result. □

Theorem 7.6.1 effectively reduces the problem of computing local Dirichlet integrals to one of treating inner and outer functions separately. We already know how to handle outer functions, so now we turn our attention to inner functions. We shall derive three formulas for the local Dirichlet integral of an inner function.

Theorem 7.6.5 *Let θ be an inner function, and let $\zeta \in \mathbb{T}$. Then*

$$\mathcal{D}_\zeta(\theta) = \lim_{r\to 1} \frac{1-|\theta(r\zeta)|^2}{1-r^2}.$$

Proof By Lemma 7.4.3,

$$\mathcal{D}_\zeta(\theta) = \lim_{r\to 1} \frac{1}{2\pi}\int_{\mathbb{T}} \frac{|\theta^*(\lambda)-\theta(r\zeta)|^2}{|\lambda-r\zeta|^2}\,|d\lambda|.$$

Now, by Lemma 7.3.3, for each $w \in \mathbb{D}$ we have

$$\begin{aligned}\frac{1}{2\pi}\int_{\mathbb{T}} \frac{|\theta^*(\lambda)-\theta(w)|^2}{|\lambda-w|^2}\,|d\lambda| &= \frac{1}{2\pi}\int_{\mathbb{T}} \frac{|\theta^*(\lambda)|^2-|\theta(w)|^2}{|\lambda-w|^2}\,|d\lambda| \\ &= \frac{1}{2\pi}\int_{\mathbb{T}} \frac{1-|\theta(w)|^2}{|\lambda-w|^2}\,|d\lambda| = \frac{1-|\theta(w)|^2}{1-|w|^2}.\end{aligned}$$

Putting these facts together, we obtain the result. □

Our second formula for $\mathcal{D}_\zeta(\theta)$ is even simpler, but supposes that ζ is a *regular point* of θ, in other words, that θ extends to be holomorphic in a neighborhood of ζ.

Theorem 7.6.6 *Let θ be an inner function, and let $\zeta \in \mathbb{T}$ be a regular point of θ. Then*

$$\mathcal{D}_\zeta(\theta) = |\theta'(\zeta)|.$$

Proof Replacing $\theta(z)$ by $\alpha\theta(\beta z)$, where α, β are suitable unimodular constants, we may suppose that $\zeta = 1$ and that $\theta(1) = 1$. Then, as $r \to 1$,

$$\frac{1-|\theta(r)|^2}{1-r^2} = \frac{1-|1+\theta'(1)(r-1)+o(r-1)|^2}{1-r^2} = \operatorname{Re}\theta'(1) + o(1-r).$$

Hence, by Theorem 7.6.5, $\mathcal{D}_1(\theta) = \operatorname{Re}\theta'(1)$. Finally, since $|\theta| = 1$ on an arc of $\mathbb{T}$ around 1, we must have $\theta'(1) \geq 0$, so $\operatorname{Re}\theta'(1) = |\theta'(1)|$. □

By Theorem A.2.4 in Appendix A, an inner function θ can be written in a unique way as Blaschke product times a singular inner function. In other words,

$$\theta(z) = c\prod_n \Big(\frac{\overline{a}_n}{|a_n|}\frac{a_n - z}{1-\overline{a}_n z}\Big)\exp\Big(-\int_{\mathbb{T}} \frac{\zeta+z}{\zeta-z}\,d\sigma(\zeta)\Big) \qquad (z\in\mathbb{D}), \tag{7.17}$$

where c is a unimodular constant, (a_n) is a sequence (finite or infinite) in $\mathbb{D}$ such that $\sum_n(1 - |a_n|) < \infty$, and σ is a finite positive Borel measure on $\mathbb{T}$ which is singular with respect to Lebesgue measure. (It is possible that a finite number of a_n are zero. Our convention is that $\overline{a}_n/|a_n| = 1$ for these terms.) Our final formula for $\mathcal{D}_\zeta(\theta)$ is given in terms of the data (a_n) and σ.

Theorem 7.6.7 *Let θ be the inner function given by* (7.17). *Then, for each $\zeta \in \mathbb{T}$,*

$$\mathcal{D}_\zeta(\theta) = \sum_n \frac{1 - |a_n|^2}{|\zeta - a_n|^2} + \int_{\mathbb{T}} \frac{2}{|\lambda - \zeta|^2}\, d\sigma(\lambda). \tag{7.18}$$

Proof We may take $c = 1$.

If $\sigma(\{\zeta\}) > 0$, then $\theta^*(\zeta) = 0$, and Theorem 7.6.5 implies that $\mathcal{D}_\zeta(\theta) = \infty$. Also clearly $\int d\sigma(\lambda)/|\lambda - \zeta| = \infty$. Thus (7.18) certainly holds in this case.

Now suppose that $\sigma(\{\zeta\}) = 0$. For $k \geq 1$, let θ_k be the inner function defined by

$$\theta_k(z) := \prod_{|a_n - \zeta| \geq 1/k} \Big(\frac{|a_n|}{a_n} \frac{a_n - z}{1 - \overline{a}_n z}\Big) \exp\Big(-\int_{|\lambda - \zeta| \geq 1/k} \frac{\lambda + z}{\lambda - z}\, d\sigma(\lambda)\Big) \quad (z \in \mathbb{D}).$$

Then ζ is a regular point of θ_k, so by the preceding theorem $\mathcal{D}_\zeta(\theta_k) = |\theta_k'(\zeta)|$. A computation using logarithmic derivatives (see Exercise 7.6.2) gives

$$|\theta_k'(\zeta)| = \sum_{|a_n - \zeta| \geq 1/k} \frac{1 - |a_n|^2}{|\zeta - a_n|^2} + \int_{|\lambda - \zeta| \geq 1/k} \frac{2}{|\lambda - \zeta|^2}\, d\sigma(\lambda). \tag{7.19}$$

As $k \to \infty$, the right-hand side tends to the right-hand side of (7.18).

It remains to show that $\mathcal{D}_\zeta(\theta_k) \to \mathcal{D}_\zeta(\theta)$ as $k \to \infty$. To see this, observe first that $\theta_k \to \theta$ locally uniformly on $\mathbb{D}$, and so by Fatou's lemma

$$\liminf_{k \to \infty} \mathcal{D}_\zeta(\theta_k) \geq \mathcal{D}_\zeta(\theta).$$

On the other hand, by Theorem 7.6.1, applied with θ_k playing the role of f and θ/θ_k playing the role of θ, we have $\mathcal{D}_\zeta(\theta) \geq \mathcal{D}_\zeta(\theta_k)$ for all k, and hence

$$\limsup_{k \to \infty} \mathcal{D}_\zeta(\theta_k) \leq \mathcal{D}_\zeta(\theta).$$

Putting these facts together gives $\mathcal{D}_\zeta(\theta_k) \to \mathcal{D}_\zeta(\theta)$ and completes the proof. □

We can now derive the formulas of Carleson and Richter–Sundberg mentioned at the beginning of the section.

Theorem 7.6.8 (Richter–Sundberg formula) *Let $f \in H^2 \setminus \{0\}$ with inner-outer factorization $f = f_i f_o$. Suppose that $f_i(z)$ is given by (7.17). Let $\zeta \in \mathbb{T}$ be a point such that $f_o^*(\zeta)$ exists. Then*

$$\mathcal{D}_\zeta(f) = I_o + I_b + I_s, \tag{7.20}$$

where

$$I_o := \frac{1}{2\pi} \int_{\mathbb{T}} \frac{|f_o^*(\lambda)|^2 - |f_o^*(\zeta)|^2 - 2|f_o^*(\zeta)|^2 \log |f_o^*(\lambda)/f_o^*(\zeta)|}{|\lambda - \zeta|^2} \, |d\lambda|,$$

$$I_b := \sum_n \frac{1 - |a_n|^2}{|\zeta - a_n|^2} |f_o^*(\zeta)|^2,$$

$$I_s := \int_{\mathbb{T}} \frac{2}{|\lambda - \zeta|^2} \, d\sigma(\lambda) |f_o^*(\zeta)|^2.$$

Proof By Theorem 7.6.1, we have

$$\mathcal{D}_\zeta(f_i f_o) = \mathcal{D}_\zeta(f_o) + |f_o^*(\zeta)|^2 \mathcal{D}_\zeta(f_i).$$

Theorem 7.4.2 gives a formula for $\mathcal{D}_\zeta(f_o)$ and Theorem 7.6.7 gives one for $\mathcal{D}_\zeta(f_i)$. The result follows. □

Theorem 7.6.9 (Carleson's formula) *Let $f \in H^2 \setminus \{0\}$ with inner-outer factorization $f = f_i f_o$. Suppose that f_i is given by (7.17). Then*

$$\mathcal{D}(f) = J_o + J_b + J_s,$$

where

$$J_o := \frac{1}{4\pi^2} \int_{\mathbb{T}} \int_{\mathbb{T}} \frac{(|f^*(\lambda)|^2 - |f^*(\zeta)|^2)(\log |f^*(\lambda)| - \log |f^*(\zeta)|)}{|\lambda - \zeta|^2} \, |d\lambda| \, |d\zeta|,$$

$$J_b := \frac{1}{2\pi} \int_{\mathbb{T}} \sum_n \frac{1 - |a_n|^2}{|\zeta - a_n|^2} |f^*(\zeta)|^2 \, |d\zeta|,$$

$$J_s := \frac{1}{2\pi} \int_{\mathbb{T}} \int_{\mathbb{T}} \frac{2}{|\lambda - \zeta|^2} |f^*(\zeta)|^2 \, d\sigma(\lambda) \, |d\zeta|.$$

Proof It suffices to integrate both sides of (7.20) with respect to normalized Lebesgue measure on $\mathbb{T}$, symmetrize as in the proof of Theorem 7.4.1, and remark that $|f_o^*| = |f^*|$ a.e. □

We conclude with a simple application that generalizes Exercise 4.1.1.

Corollary 7.6.10 *The only inner functions in $\mathcal{D}$ are finite Blaschke products.*

Proof Applying Theorem 7.6.9 with $f = f_i$ and $f_o = 1$ gives

$$\mathcal{D}(f_i) = \frac{1}{2\pi}\int_{\mathbb{T}}\sum_n \frac{1-|a_n|^2}{|\zeta - a_n|^2}\,|d\zeta| + \frac{1}{2\pi}\int_{\mathbb{T}}\int_{\mathbb{T}}\frac{2}{|\lambda-\zeta|^2}\,d\sigma(\lambda)\,|d\zeta|.$$

Now

$$\int_{\mathbb{T}}\frac{2}{|\lambda-\zeta|^2}\,|d\zeta| = \infty \qquad (\lambda \in \mathbb{T}),$$

so $\mathcal{D}(f_i)$ can only be finite if $\sigma = 0$. Also

$$\frac{1}{2\pi}\int_{\mathbb{T}}\frac{1-|w|^2}{|\zeta-w|^2}\,|d\zeta| = 1 \qquad (w \in \mathbb{D}),$$

so $\mathcal{D}(f_i)$ can only be finite if the set of zeros (a_n) is finite. In this case, f_i is a finite Blaschke product, and $\mathcal{D}(f_i)$ is equal to the degree of f_i. □

Exercises 7.6

1. Let $f \in H^2$, let θ be an inner function, and suppose that $\theta f \in \mathcal{D}_\mu$. Show that, for each $n \geq 1$, we have $\theta^n f \in \mathcal{D}_\mu$ and that $\mathcal{D}_\mu(\theta^n f) \leq n\mathcal{D}_\mu(\theta f)$.
2. Let θ_1, θ_2 be inner functions. Suppose that $\zeta \in \mathbb{T}$ is a regular point for both θ_1, θ_2. Prove that
$$|(\theta_1\theta_2)'(\zeta)| = |\theta_1'(\zeta)| + |\theta_2'(\zeta)|.$$
Use this to prove formula (7.19).
3. Let θ be the inner function defined by
$$\theta(z) := \exp\Big(\frac{z-1}{z+1}\Big) \qquad (z \in \mathbb{D}),$$
and set $f := 1/\theta$. The formula (7.16) clearly fails in this case. Where does the proof break down?
4. Using Theorem 7.4.5, show that Theorems 7.6.8 and 7.6.9 hold for all functions of the form $f = f_i f_o$, where f_i is inner and f_o is outer, irrespective of whether $f \in H^2$. [This set of functions is the so-called *Smirnov class*; see Appendix A.3.]

Notes on Chapter 7

§7.1

The $\mathcal{D}_\mu$ spaces were introduced by Richter in [96], as part of his analysis of 2-isometries (more on this in the next chapter). Aleman [7] introduced a more general class of Dirichlet spaces with positive superharmonic weights, which includes both the $\mathcal{D}_\mu$ spaces and the $\mathcal{D}_\alpha$ spaces $(0 < \alpha < 1)$ mentioned in §1.6.

§7.2

The local Dirichlet integral was studied systematically by Richter and Sundberg in their article [98], on which a large part of this chapter is based. In particular, they established the local Douglas formula (7.4). In fact, they took this formula as the definition of local Dirichlet integral, showing subsequently that it is indeed equivalent to the one given given earlier.

§7.3

The main approximation theorem 7.3.1 was obtained by Richter in [96]. He established a weaker version of Lemma 7.3.2, in which $\mathcal{D}_\mu(f_r) \leq C\mathcal{D}_\mu(f)$ for some unknown constant C. Richter and Sundberg [98] showed that this inequality holds with $C = 4$, and Aleman [7] improved this to $C = 5/2$. Finally Sarason [108] gave a proof with $C = 1$ by identifying the local Dirichlet spaces $\mathcal{D}_\zeta$ as a special case of so-called de Branges–Rovnyak spaces. The proof given here is an adaptation of his idea.

§7.4

Theorem 7.4.1 was originally obtained (with a different proof) by Carleson [27]. The local version, Theorem 7.4.2, is due to Richter and Sundberg [98].

§7.5

The results of this section are taken from the paper [99] of Richter and Sundberg.

§7.6

Theorem 7.6.9 was originally obtained (with a different proof) by Carleson [27]. The local version, Theorem 7.6.8, together with the other theorems of this section, are due to Richter and Sundberg [98]. Sarason had already obtained a precursor of Theorem 7.6.5 in [107], and had remarked on the close connection with angular derivatives and the Julia–Carathéodory theorem.

Finally, we remark that several of the themes already studied for $\mathcal{D}$ have analogues for $\mathcal{D}_\mu$ that are the subjects of recent research. These include boundary behavior [53], zero sets [52], Carleson measures [33, 35] and Pick interpolation [116].

8

Invariant subspaces

'Invariant subspaces' here refers to the closed subspaces invariant under the shift operator $f \mapsto zf$. The invariant subspaces of H^2 were completely classified by Beurling, who showed that they are of the form θH^2 for some inner function θ (together with the zero subspace). It is no exaggeration to say that Beurling's theorem is one of the cornerstones of the whole theory of Hardy spaces. Our aim in this chapter is to obtain an analogous result in the Dirichlet space $\mathcal{D}$.

One of the principal differences between H^2 and $\mathcal{D}$ is that, whereas the shift operator acts as an isometry on H^2, this is no longer true on $\mathcal{D}$. Instead, it is a so-called 2-isometry. One of our main tasks will be to study 2-isometries in general, and to prove a representation theorem for them. This theorem brings into play the $\mathcal{D}_\mu$-spaces introduced in the previous chapter. We shall then exploit the representation theorem to study the structure of invariant subspaces of $\mathcal{D}$ and (because it involves virtually no extra work) those of $\mathcal{D}_\mu$.

8.1 The shift operator on $\mathcal{D}_\mu$

Let μ be a finite measure on $\mathbb{T}$, and let $\mathcal{D}_\mu$ be the corresponding harmonically weighted Dirichlet space defined in the previous chapter.

Definition 8.1.1 The *shift operator* on $\mathcal{D}_\mu$ is the operator M_z defined by

$$(M_z f)(z) := zf(z) \qquad (f \in \mathcal{D}_\mu).$$

Our first task is to show that M_z is indeed a bounded operator on $\mathcal{D}_\mu$.

Theorem 8.1.2 *Let $f \in \mathcal{D}_\mu$. Then the following statements hold.*

(i) *The radial limit $f^*(\zeta)$ exists μ-almost everywhere on $\mathbb{T}$.*

(ii) $f^* \in L^2(\mu)$ *and* $\|f^*\|_{L^2(\mu)} \le (1 + \mu(\mathbb{T})^{1/2})\|f\|_{\mathcal{D}_\mu}$.

(iii) $zf \in \mathcal{D}_\mu$ *and*

$$\|zf\|^2_{\mathcal{D}_\mu} = \|f\|^2_{\mathcal{D}_\mu} + \|f^*\|^2_{L^2(\mu)}. \tag{8.1}$$

Proof (i) This was already proved in Corollary 7.2.4.

(ii) Define $g(z) := (f(z) - f(0))/z$. Since $f \in H^2$, we also have $g \in H^2$. Thus we can apply Theorem 7.6.1 to g, with $\theta(z) = z$, to obtain

$$\mathcal{D}_\zeta(zg) = \mathcal{D}_\zeta(g) + |g^*(\zeta)|^2 \mathcal{D}_\zeta(z).$$

As $\mathcal{D}_\zeta(z) = 1$ and $f = zg + f(0)$, this is equivalent to

$$\mathcal{D}_\zeta(f) = \mathcal{D}_\zeta(g) + |f^*(\zeta) - f(0)|^2.$$

Integrating with both sides respect to μ, we deduce that

$$\int_{\mathbb{T}} |f^*(\zeta) - f(0)|^2 \, d\mu(\zeta) \le \mathcal{D}_\mu(f).$$

Hence $f^* \in L^2(\mu)$ and

$$\|f^*\|_{L^2(\mu)} \le \mathcal{D}_\mu(f)^{1/2} + |f(0)|\mu(\mathbb{T})^{1/2} \le (1 + \mu(\mathbb{T})^{1/2})\|f\|_{\mathcal{D}_\mu}.$$

(iii) Re-apply Theorem 7.6.1 directly to f, still with $\theta(z) = z$, to obtain

$$\mathcal{D}_\zeta(zf) = \mathcal{D}_\zeta(f) + |f^*(\zeta)|^2.$$

Integrating both sides of this equation with respect to μ gives

$$\mathcal{D}_\mu(zf) = \mathcal{D}_\mu(f) + \|f^*\|^2_{L^2(\mu)},$$

which is equivalent to (8.1). In particular, $\mathcal{D}_\mu(zf) < \infty$. □

Corollary 8.1.3 *M_z is a bounded operator on $\mathcal{D}_\mu$ with $\|M_z\| \le \sqrt{2} + \mu(\mathbb{T})^{1/2}$.*

Proof By parts (ii) and (iii) of the theorem,

$$\|zf\|^2_{\mathcal{D}_\mu} \le (1 + (1 + \mu(\mathbb{T})^{1/2})^2)\|f\|^2_{\mathcal{D}_\mu} \le (\sqrt{2} + \mu(\mathbb{T})^{1/2})^2\|f\|^2_{\mathcal{D}_\mu}.$$

This is equivalent to the statement of the corollary. □

In the light of this result, it is natural to ask when $M_\phi : f \mapsto \phi f$ is a bounded operator on $\mathcal{D}_\mu$. This is equivalent to asking whether ϕ is a multiplier on $\mathcal{D}_\mu$. We shall not pursue this question in the same depth as we did for $\mathcal{D}$ in Chapter 5, contenting ourselves with the following sufficient condition.

Theorem 8.1.4 *Let $f \in \mathcal{D}_\mu$ and let $\phi \in \mathrm{Hol}(\mathbb{D})$ with $\phi' \in H^\infty$. Then $\phi f \in \mathcal{D}_\mu$, and*

$$\mathcal{D}_\mu(\phi f) \le 2\|\phi\|^2_{H^\infty}\mathcal{D}_\mu(f) + 2\|\phi'\|^2_{H^\infty}\|f^*\|^2_{L^2(\mu)}. \tag{8.2}$$

The proof depends on a technical estimate.

Lemma 8.1.5 *Let $f \in H^2$ and let $\phi \in H^\infty$. Let $\zeta \in \mathbb{T}$ be a point such that $f^*(\zeta)$ exists. Then*

$$\mathcal{D}_\zeta(\phi f) \le 2\|\phi\|_{H^\infty}^2 \mathcal{D}_\zeta(f) + 2\mathcal{D}_\zeta(\phi)|f^*(\zeta)|^2,$$

and

$$\mathcal{D}_\zeta(\phi)|f^*(\zeta)|^2 \le 2\|\phi\|_{H^\infty}^2 \mathcal{D}_\zeta(f) + 2\mathcal{D}_\zeta(\phi f).$$

Proof If $f^*(\zeta) = 0$, then clearly $\mathcal{D}_\zeta(\phi f) \le \|\phi\|_{H^\infty}^2 \mathcal{D}_\zeta(f)$. So we may as well suppose that $f^*(\zeta) \ne 0$ and that $\mathcal{D}_\zeta(\phi) < \infty$, which implies that $\phi^*(\zeta)$ exists. We then have

$$\begin{aligned}
\mathcal{D}_\zeta(\phi f) &= \frac{1}{2\pi}\int_{\mathbb{T}} \frac{|\phi^*(\lambda)f^*(\lambda) - \phi^*(\zeta)f^*(\zeta)|^2}{|\lambda - \zeta|^2}\,|d\lambda| \\
&\le \frac{1}{2\pi}\int_{\mathbb{T}} \Big(2|\phi^*(\lambda)|^2 \frac{|f^*(\lambda) - f^*(\zeta)|^2}{|\lambda - \zeta|^2} + 2|f^*(\zeta)|^2 \frac{|\phi^*(\lambda) - \phi^*(\zeta)|^2}{|\lambda - \zeta|^2}\Big)|d\lambda| \\
&\le 2\|\phi\|_{H^\infty}^2 \mathcal{D}_\zeta(f) + 2|f^*(\zeta)|^2 \mathcal{D}_\zeta(\phi).
\end{aligned}$$

This gives the first inequality. The second inequality is proved the same way. □

Proof of Theorem 8.1.4 Since $\phi' \in H^\infty$, we have $\mathcal{D}_\zeta(\phi) \le \|\phi'\|_{H^\infty}^2$ for all $\zeta \in \mathbb{T}$ (see Exercise 7.1.2). Clearly also $\phi \in H^\infty$. By Lemma 8.1.5, it follows that, for μ-almost every $\zeta \in \mathbb{T}$,

$$\mathcal{D}_\zeta(\phi f) \le 2\|\phi\|_{H^\infty}^2 \mathcal{D}_\zeta(f) + 2\|\phi'\|_{H^\infty}^2 |f^*(\zeta)|^2.$$

Integrating both sides with respect to $d\mu(\zeta)$, we get (8.2). □

This theorem does not extend to multiplication by H^∞-functions or even by inner functions. Indeed, we have already seen that the only inner functions in $\mathcal{D}$ are finite Blaschke products (Corollary 7.6.10). However, it is always possible to *divide* by inner functions, in the sense of Corollary 7.6.2. This will prove useful at several points later in the chapter.

Exercises 8.1

1. Let $S^* : H^2 \to H^2$ be the adjoint of the shift operator on H^2, namely $S^*f(z) := (f(z) - f(0))/z$. Let $\zeta \in \mathbb{T}$ and let $f \in \mathcal{D}_\zeta$.
   (i) Show that $S^*f \in \mathcal{D}_\zeta$ and that $\|S^*f\|_{\mathcal{D}_\zeta} \le \|f\|_{\mathcal{D}_\zeta}$.
   (ii) Show that $\|S^{*n}f\|_{\mathcal{D}_\zeta} \to 0$ as $n \to \infty$. [Hint: Use Corollary 7.3.4.]
   (iii) Show that $\mathcal{D}_\zeta(f) = \mathcal{D}_\zeta(S^*f) + |(S^*f)^*(\zeta)|^2$.

(iv) Deduce that

$$\mathcal{D}_\zeta(f) = \sum_{k\geq 1} |(S^{*k}f)^*(\zeta)|^2.$$

[Compare this formula with the one obtained in Exercise 7.2.4.]

8.2 Characterization of the shift operator

We begin this section by establishing some further properties of $(M_z, \mathcal{D}_\mu)$.

Theorem 8.2.1 *$(M_z, \mathcal{D}_\mu)$ satisfies the following.*

(i) $\|M_z^2 f\|^2_{\mathcal{D}_\mu} - 2\|M_z f\|^2_{\mathcal{D}_\mu} + \|f\|^2_{\mathcal{D}_\mu} = 0$ *for all* $f \in \mathcal{D}_\mu$.
(ii) $\cap_{n\geq 0} M_z^n(\mathcal{D}_\mu) = \{0\}$.
(iii) $\dim(\mathcal{D}_\mu \ominus M_z(\mathcal{D}_\mu)) = 1$.

Proof (i) Using (8.1), we have

$$\|z^2 f\|^2_{\mathcal{D}_\mu} - \|zf\|^2_{\mathcal{D}_\mu} = \|zf^*\|^2_{L^2(\mu)} = \|f^*\|^2_{L^2(\mu)} = \|zf\|^2_{\mathcal{D}_\mu} - \|f\|^2_{\mathcal{D}_\mu}.$$

(ii) If $f \in \cap_{n\geq 0} M_z^n(\mathcal{D}_\mu)$, then f is holomorphic on $\mathbb{D}$ with a zero of infinite order at the origin, so $f = 0$.

(iii) If $f \in \mathcal{D}_\mu$ and $f(0) = 0$, then $f = zg$, where $g \in \mathcal{D}_\mu$ by Corollary 7.6.2. Thus $M_z(\mathcal{D}_\mu)$ is precisely the set of $f \in \mathcal{D}_\mu$ such that $f(0) = 0$, which is a closed subspace of codimension one. □

The three properties of Theorem 8.2.1 were singled out because they completely characterize $(M_z, \mathcal{D}_\mu)$. This is the conclusion of the following theorem.

Theorem 8.2.2 *Let T be a bounded operator on a Hilbert space $\mathcal{H}$ such that:*

(i) $\|T^2 x\|^2 - 2\|Tx\|^2 + \|x\|^2 = 0$ *for all* $x \in \mathcal{H}$;
(ii) $\cap_{n\geq 0} T^n(\mathcal{H}) = \{0\}$;
(iii) $\dim(\mathcal{H} \ominus T(\mathcal{H})) = 1$.

Then there exists a unique finite measure μ on $\mathbb{T}$ such that $(T, \mathcal{H})$ is unitarily equivalent to $(M_z, \mathcal{D}_\mu)$.

An operator T that satisfies property (i) is called a *2-isometry*, and one that satisfies property (ii) is called *analytic*. We shall see in the course of the proof that, under the conditions (i) and (ii), property (iii) is equivalent to T being *cyclic* (i.e., there exists a vector x such that $\{T^n x : n \geq 0\}$ spans a dense subspace of $\mathcal{H}$). Thus Theorem 8.2.2 is sometimes stated in the form that every cyclic, analytic, 2-isometry is unitarily equivalent to $(M_z, \mathcal{D}_\mu)$ for some μ.

The proof of Theorem 8.2.2 will take up the rest of the section. The first step is to establish an elementary but important property of 2-isometries.

Lemma 8.2.3 *If $(T, \mathcal{H})$ is a 2-isometry, then $\|Tx\| \geq \|x\|$ for all $x \in \mathcal{H}$. Consequently, the image of each closed subspace of $\mathcal{H}$ under T is again a closed subspace of $\mathcal{H}$.*

Proof As T is a 2-isometry,

$$\|T^2x\|^2 - \|Tx\|^2 = \|Tx\|^2 - \|x\|^2 \qquad (x \in \mathcal{H}).$$

Replacing x by T^jx for $j = 0, \ldots, k-2$, we find that

$$\|T^kx\|^2 - \|T^{k-1}x\|^2 = \|Tx\|^2 - \|x\|^2 \qquad (k \geq 1).$$

Therefore, if we sum from $k = 1$ to n, we obtain

$$\|T^nx\|^2 - \|x\|^2 = n(\|Tx\|^2 - \|x\|^2) \qquad (n \geq 1). \tag{8.3}$$

In particular, $\|Tx\|^2 - \|x\|^2 \geq -\|x\|^2/n$. Letting $n \to \infty$ gives the result. □

Theorem 8.2.4 (Wandering subspace theorem) *Let $(T, \mathcal{H})$ be an analytic 2-isometry. Then $\{T^n(\mathcal{H} \ominus T(\mathcal{H})) : n \geq 0\}$ spans a dense subspace of $\mathcal{H}$.*

Proof By Lemma 8.2.3, the operator $T^*T - I$ is positive. Therefore T^*T is invertible, and $L := (T^*T)^{-1}T^*$ is a left inverse of T. The operator $Q := TL$ is the orthogonal projection onto $T(\mathcal{H})$, and $P := I - Q$ is the orthogonal projection onto $\mathcal{H} \ominus T(\mathcal{H})$. Note also that L is a contraction: indeed, using Lemma 8.2.3 again, we have

$$\|Lx\| \leq \|TLx\| = \|Qx\| \leq \|x\| \quad (x \in \mathcal{H}).$$

Let $x \in \mathcal{H}$. Then, for each $n \geq 1$,

$$x - T^nL^nx = \sum_{k=0}^{n-1}(T^kL^kx - T^{k+1}L^{k+1}x) = \sum_{k=0}^{n-1} T^kPL^kx \in \sum_{k=0}^{n-1} T^kP(\mathcal{H}).$$

The right-hand side lies in the span of $\{T^k(\mathcal{H} \ominus T(\mathcal{H})) : k \geq 0\}$. Thus it suffices to show that some subsequence of $(T^nL^nx)_{n\geq 0}$ converges weakly to zero.

Replacing x by L^nx in (8.3), we have

$$\begin{aligned}\|T^nL^nx\|^2 - \|L^nx\|^2 &= n(\|TL^nx\|^2 - \|L^nx\|^2)\\ &= n(\|QL^{n-1}x\|^2 - \|L^nx\|^2)\\ &\leq n(\|L^{n-1}x\|^2 - \|L^nx\|^2).\end{aligned}$$

Divide by n and sum:

$$\sum_{n\geq 1}\frac{\|T^nL^nx\|^2-\|L^nx\|^2}{n}\leq \|x\|^2<\infty.$$

We infer that $\liminf_{n\to\infty}(\|T^nL^nx\|^2-\|L^nx\|^2)=0$. Also, as L is a contraction, the sequence (L^nx) is bounded. Therefore some subsequence of (T^nL^nx) is bounded, and so some further subsequence of (T^nL^nx) converges weakly to an element $y\in\mathcal{H}$. As the sequence $(T^n\mathcal{H})_{n\geq 0}$ is decreasing, it follows that $y\in\cap_{n\geq 0}T^n\mathcal{H}$. By assumption, T is analytic, so $\cap_{n\geq 0}T^n\mathcal{H}=\{0\}$. Therefore $y=0$, and the proof is complete. □

Corollary 8.2.5 *Let $(T,\mathcal{H})$ be an analytic 2-isometry. Then T is cyclic if and only if $\dim(\mathcal{H}\ominus T(\mathcal{H}))=1$. In this case, each non-zero vector of $\mathcal{H}\ominus T(\mathcal{H})$ is cyclic for T.*

Proof If $\dim(\mathcal{H}\ominus T(\mathcal{H}))=1$, then by the theorem it is clear that each non-zero vector of $\mathcal{H}\ominus T(\mathcal{H})$ is cyclic for T.

Conversely, it is obvious that, if T has a cyclic vector x, then $\mathcal{H}$ is spanned by $T(\mathcal{H})\cup\{x\}$, so $\dim(\mathcal{H}\ominus T(\mathcal{H}))\leq 1$. Moreover, $T(\mathcal{H})$ is a proper, closed subspace of $\mathcal{H}$, so in fact $\dim(\mathcal{H}\ominus T(\mathcal{H}))=1$. □

Applying Corollary 8.2.5 to $(M_z,\mathcal{D}_\mu)$, we obtain that 1 is a cyclic vector, thereby recovering the fact that polynomials are dense in $\mathcal{D}_\mu$ (Corollary 7.3.4). However, this is not really a new proof, because the density of polynomials in $\mathcal{D}_\mu$ is used implicitly in showing that $(M_z,\mathcal{D}_\mu)$ is a 2-isometry.

The next step is to construct the measure μ that appears in Theorem 8.2.2.

Theorem 8.2.6 *Let $(T,\mathcal{H})$ be a 2-isometry and let $a\in\mathcal{H}\ominus T(\mathcal{H})$ be an element of norm 1. Then there exists a unique positive finite Borel measure μ on $\mathbb{T}$ such that, for all polynomials p,*

$$\|p(T)a\|=\|p\|_{\mathcal{D}_\mu}.$$

For the proof, we need a lemma.

Lemma 8.2.7 *Let $(T,\mathcal{H})$ be a 2-isometry and let $a\in\mathcal{H}$. Then there exists a positive finite Borel measure μ on $\mathbb{T}$ such that, for every polynomial p,*

$$\|Tp(T)a\|^2-\|p(T)a\|^2=\int_{\mathbb{T}}|p|^2\,d\mu.$$

Proof Denote by $\langle\cdot,\cdot\rangle$ the inner product on $\mathcal{H}$, and let us define a second sesquilinear form on $\mathcal{H}$ by the formula

$$[x,y]:=\langle Tx,Ty\rangle-\langle x,y\rangle \qquad (x,y\in\mathcal{H}).$$

Note that $[x, x] = \|Tx\|^2 - \|x\|^2 \geq 0$ for all x, by Lemma 8.2.3, so $[\cdot, \cdot]$ is a semi-inner product on $\mathcal{H}$. Further, the fact that T is a 2-isometry translates to $[Tx, Tx] = [x, x]$, in other words, T is an isometry with respect to $[\cdot, \cdot]$. By polarization, it follows that $[Tx, Ty] = [x, y]$ for all $x, y \in \mathcal{H}$.

Define a linear functional Λ on the trigonometric polynomials by stipulating that

$$\Lambda(e^{in\theta}) = \overline{\Lambda(e^{-in\theta})} := [T^n a, a] \qquad (n \geq 0).$$

From the property that $[Tx, Ty] = [x, y]$, it then follows that

$$\Lambda(e^{i(j-k)\theta}) = [T^j a, T^k a] \qquad (j, k \geq 0).$$

We claim that Λ is a positive linear functional. Indeed, by the Fejér–Riesz theorem, every non-negative trigonometric polynomial is of the form $|p|^2$ for some analytic polynomial p, and if $p(z) = \sum_{j=0}^n c_j z^j$, then

$$\Lambda(|p|^2) = \Lambda\Big(\sum_{j,k} c_j \bar{c}_k e^{i(j-k)\theta}\Big) = \sum_{j,k} c_j \bar{c}_k [T^j a, T^k a] = [p(T)a, p(T)a] \geq 0.$$

The claim is justified.

As the trigonometric polynomials are uniformly dense in $C(\mathbb{T})$, there is a (unique) extension of Λ to a positive linear functional on $C(\mathbb{T})$. By the Riesz representation theorem for continuous functions, there is a (unique) positive finite Borel measure μ on $\mathbb{T}$ such that $\Lambda(\phi) = \int \phi \, d\mu$ for all $\phi \in C(\mathbb{T})$. In particular, if p is an analytic polynomial, then

$$\int |p|^2 \, d\mu = \Lambda(|p|^2) = [p(T)a, p(T)a] = \|Tp(T)a\|^2 - \|p(T)a\|^2.$$

Thus μ has the required property, and the proof of the lemma is complete. □

Proof of Theorem 8.2.6 First we prove uniqueness. Let μ and ν be two finite Borel measures on $\mathbb{T}$ satisfying the conclusion of the theorem. Then we have $\|p\|_{\mathcal{D}_\mu} = \|p\|_{\mathcal{D}_\nu}$ for every polynomial p. Using (8.1), we deduce that, for every polynomial p,

$$\int_{\mathbb{T}} |p|^2 \, d\mu = \int_{\mathbb{T}} |p|^2 \, d\nu.$$

By polarization, it follows that, for every pair of polynomials p, q,

$$\int_{\mathbb{T}} p\bar{q} \, d\mu = \int_{\mathbb{T}} p\bar{q} \, d\nu.$$

In particular, taking $p(z) = z^n$ and $q(z) = z^m$, we find that μ and ν have exactly the same Fourier coefficients, and consequently $\mu = \nu$.

Now we turn to the proof of existence. By Lemma 8.2.7, there exists a finite positive measure μ on $\mathbb{T}$ such that, for every polynomial p,

$$\|zp\|_{\mathcal{D}_\mu}^2 - \|p\|_{\mathcal{D}_\mu}^2 = \int_{\mathbb{T}} |p|^2 \, d\mu = \|Tp(T)a\|^2 - \|p(T)a\|^2. \tag{8.4}$$

Define a sesquilinear form on polynomials by

$$(p, q) := \langle p(T)a, q(T)a \rangle - \langle p, q \rangle_{\mathcal{D}_\mu}.$$

Equation (8.4) translates to $(zp, zp) = (p, p)$, which, by polarization, implies that

$$(zp, zq) = (p, q) \quad \text{for all polynomials } p, q. \tag{8.5}$$

Also, since $a \in \mathcal{H} \ominus T(\mathcal{H})$, we have $\langle a, T^n a \rangle = 0$ for all $n \geq 1$, and for $n = 0$ we have $\langle a, a \rangle = \|a\|^2 = 1$. Together, these imply

$$(1, z^n) = 0 \qquad (n \geq 0). \tag{8.6}$$

Combining (8.5) and (8.6), we see that $(z^n, z^m) = 0$ for all $m, n \geq 0$, and hence that $(p, q) = 0$ for all polynomials p, q. In particular, $(p, p) = 0$, which amounts to saying that $\|p(T)a\|^2 = \|p\|_{\mathcal{D}_\mu}^2$, as desired. □

Finally, we have all the ingredients necessary to prove Theorem 8.2.2.

Proof of Theorem 8.2.2 We begin with the uniqueness of μ. The conclusion of the theorem is that there is a unitary operator $U : \mathcal{D}_\mu \to \mathcal{H}$ such that $U^{-1}TU = M_z$. Then $p(T)U = Up(M_z)$ for every polynomial p, and in particular $p(T)U(1) = Up(M_z)(1) = U(p)$. Thus, writing $a := U(1)$, we have $a \in \mathcal{H} \ominus T(\mathcal{H})$ and $p(T)a = U(p)$. In particular, $\|p(T)a\| = \|p\|_{\mathcal{D}_\mu}$. Therefore the uniqueness of μ follows from the uniqueness part of Theorem 8.2.6.

Now we turn the existence of μ. By assumption $\dim(\mathcal{H} \ominus T(\mathcal{H})) = 1$. Pick $a \in \mathcal{H} \ominus T(\mathcal{H})$ with $\|a\| = 1$. By Theorem 8.2.6, there exists a unique positive finite measure μ on $\mathbb{T}$ such that $\|p(T)a\| = \|p\|_{\mathcal{D}_\mu}$ for all polynomials p. Thus, if we define $U(p) := p(T)a$, then U is an isometry from the space of polynomials, thought of as a subspace of $\mathcal{D}_\mu$, into the Hilbert space $\mathcal{H}$. As the polynomials are dense in $\mathcal{D}_\mu$, the operator U extends to an isometry of the whole of $\mathcal{D}_\mu$ onto a closed subspace of $\mathcal{H}$. Moreover, by Corollary 8.2.5, the vector a is a cyclic vector for T, which implies that the range of U is dense in $\mathcal{H}$. Thus in fact U is a surjective isometry, and therefore a unitary operator. Finally, since $U^{-1}TU(p) = zp$, it follows that $U^{-1}TU = M_z$. Hence $(T, \mathcal{H})$ is unitarily equivalent to $(M_z, \mathcal{D}_\mu)$. □

Exercises 8.2

1. Let $(T, \mathcal{H})$ be a 2-isometry that is not an isometry. Show that $\|T^n\| \asymp \sqrt{n}$ as $n \to \infty$. [Hint: Use formula (8.3).]
2. Let $(T, \mathcal{H})$ be a 2-isometry. Set $\mathcal{H}_1 := \cap_{n \geq 0} T^n(\mathcal{H})$ and $\mathcal{H}_2 := \mathcal{H} \ominus \mathcal{H}_1$.
   (i) Show that $T(\mathcal{H}_1) = \mathcal{H}_1$, and deduce that $T|_{\mathcal{H}_1}$ is invertible.
   (ii) Show that $(T|_{\mathcal{H}_1})^{-1}$ is a 2-isometry
   (iii) Deduce that $T|_{\mathcal{H}_1}$ is unitary.
   (iv) Show that $T(\mathcal{H}_2) \subset \mathcal{H}_2$.
   (v) Conclude that $T = U \oplus S$, where U is unitary on $\mathcal{H}_1$ and S is an analytic 2-isometry on $\mathcal{H}_2$.

8.3 Invariant subspaces of $\mathcal{D}_\mu$

Let μ be a finite measure on $\mathbb{T}$. Our goal in this section is to describe the closed shift-invariant subspaces of $\mathcal{D}_\mu$. It is convenient to introduce a little notation.

Definition 8.3.1 Given a bounded operator T on a Hilbert space $\mathcal{H}$, we write $\mathrm{Lat}(T, \mathcal{H})$ for the family of closed subspaces $\mathcal{M}$ of $\mathcal{H}$ such that $T(\mathcal{M}) \subset \mathcal{M}$.

The elements of $\mathrm{Lat}(T, \mathcal{H})$ form a lattice under the operations $\mathcal{M} \cap \mathcal{N}$ and $\overline{\mathcal{M} + \mathcal{N}}$ (whence the notation).

We seek to describe $\mathrm{Lat}(M_z, \mathcal{D}_\mu)$. From the very definition, it is clear that if $\mathcal{M} \in \mathrm{Lat}(M_z, \mathcal{D}_\mu)$, then $p\mathcal{M} \subset \mathcal{M}$ for every polynomial p. We extend this observation to more general functions.

Theorem 8.3.2 *Let $\mathcal{M} \in \mathrm{Lat}(M_z, \mathcal{D}_\mu)$ and let $\phi \in H^\infty$. Then $\phi\mathcal{M} \cap \mathcal{D}_\mu \subset \mathcal{M}$.*

Proof Let $f \in \mathcal{M}$ and suppose that $\phi f \in \mathcal{D}_\mu$. We need to show that $\phi f \in \mathcal{M}$.

Suppose first that ϕ is holomorphic in a neighborhood of $\overline{\mathbb{D}}$. Let ϕ_n be the n-th partial sum of the Taylor series of ϕ. Clearly $\phi_n f \in \mathcal{M}$ for every n. Also $\phi_n \to \phi$ and $\phi_n' \to \phi'$ uniformly on $\mathbb{D}$, so by Theorem 8.1.4, $\|\phi_n f - \phi f\|_{\mathcal{D}_\mu} \to 0$. Hence $\phi f \in \mathcal{M}$ in this case.

Now suppose merely that $\phi \in H^\infty$. For $0 < r < 1$, we consider $\phi_r(z) := \phi(rz)$. Then ϕ_r is holomorphic in a neighborhood of $\overline{\mathbb{D}}$, so, by what we just proved, $\phi_r f \in \mathcal{M}$. We claim that $\sup_{r<1} \mathcal{D}_\mu(\phi_r f) < \infty$. If this is indeed the case, then the family $(\phi_r f)$ is bounded in $\mathcal{D}_\mu$, so there exists a sequence $r_n \to 1^-$ along which $(\phi_{r_n} f)$ converges weakly in $\mathcal{D}_\mu$, necessarily to ϕf. As $\mathcal{M}$ is a closed subspace of $\mathcal{D}_\mu$, it is weakly closed, and consequently $\phi f \in \mathcal{M}$, as desired.

It remains to justify the claim that $\sup_{r<1} \mathcal{D}_\mu(\phi_r f) < \infty$. To see this, we use Lemma 8.1.5, Lemma 7.3.2 and Lemma 8.1.5 again to obtain, for each $\zeta \in \mathbb{T}$,

$$\begin{aligned}\mathcal{D}_\zeta(\phi_r f) &\leq 2\|\phi_r\|_{H^\infty}^2 \mathcal{D}_\zeta(f) + 2\mathcal{D}_\zeta(\phi_r)|f^*(\zeta)|^2 \\ &\leq 2\|\phi\|_{H^\infty}^2 \mathcal{D}_\zeta(f) + 2\mathcal{D}_\zeta(\phi)|f^*(\zeta)|^2 \\ &\leq 6\|\phi\|_{H^\infty}^2 \mathcal{D}_\zeta(f) + 4\mathcal{D}_\zeta(\phi f).\end{aligned}$$

Integrating with respect to $d\mu(\zeta)$ gives

$$\mathcal{D}_\mu(\phi_r f) \leq 6\|\phi\|_{H^\infty}^2 \mathcal{D}_\mu(f) + 4\mathcal{D}_\mu(\phi f).$$

In particular, $\sup_{r<1} \mathcal{D}_\mu(\phi_r f) < \infty$, justifying the claim. □

Corollary 8.3.3 *Let $\mathcal{M} \in \mathrm{Lat}(M_z, \mathcal{D}_\mu)$, let $f \in \mathcal{M}$, and let $g \in \mathcal{D}_\mu$ with $|g| \leq |f|$. Then $g \in \mathcal{M}$.*

Proof Apply the theorem with $\phi := g/f$. □

The next result is a density theorem for invariant subspaces.

Theorem 8.3.4 *Let $\mathcal{M} \in \mathrm{Lat}(M_z, \mathcal{D}_\mu)$. Then $\mathcal{M} \cap H^\infty$ is dense in $\mathcal{M}$.*

The proof is based upon an approximation lemma.

Lemma 8.3.5 *Let $f \in \mathcal{D}_\mu$. Then there exists a sequence (f_n) in $\mathcal{D}_\mu \cap H^\infty$ such that $|f_n| \leq |f|$ for all n and $\|f_n - f\|_{\mathcal{D}_\mu} \to 0$ as $n \to \infty$.*

Proof Let $f = f_i f_o$ be the inner-outer factorization of f. Consider $g_n := f_i(f_o \wedge n)$. By Corollary 7.6.4, we have $g_n \in \mathcal{D}_\mu$ with $\mathcal{D}_\mu(g_n) \leq \mathcal{D}_\mu(f)$. Hence the sequence (g_n) is bounded in $\mathcal{D}_\mu$. Therefore some subsequence converges weakly. As the whole sequence converges pointwise to f, the weak limit of the subsequence is f. By taking appropriate convex combinations of members of this subsequence, we can obtain a sequence (f_n) which converges in norm to f. Further, since the g_n belong to H^∞ and satisfy $|g_n| \leq |f|$, the same is clearly true of the f_n. □

Proof of Theorem 8.3.4 Let $f \in \mathcal{M}$. By the lemma there exists a sequence (f_n) in $\mathcal{D}_\mu \cap H^\infty$ such that $\|f_n - f\|_{\mathcal{D}_\mu} \to 0$ and $|f_n| \leq |f|$. By Corollary 8.3.3, the latter inequality implies that $f_n \in \mathcal{M}$ for all n. Thus f is the limit of functions in $\mathcal{M} \cap H^\infty$. □

We saw in Theorem 8.2.1 that $(M_z, \mathcal{D}_\mu)$ is an analytic 2-isometry. Evidently the same is true for the restriction $(M_z, \mathcal{M})$ for each $\mathcal{M} \in \mathrm{Lat}(M_z, \mathcal{D}_\mu)$. Thus $(M_z, \mathcal{M})$ satisfies the hypotheses (i) and (ii) of Theorem 8.2.2. We now show that it also satisfies hypothesis (iii).

Theorem 8.3.6 *Let $\mathcal{M} \in \text{Lat}(M_z, \mathcal{D}_\mu)$, $\mathcal{M} \neq \{0\}$. Then* $\dim(\mathcal{M} \ominus z\mathcal{M}) = 1$.

Proof Since $\mathcal{M} \neq \{0\}$, the space $z\mathcal{M}$ is a proper closed subspace of $\mathcal{M}$, so clearly $\dim(\mathcal{M} \ominus z\mathcal{M}) \geq 1$. The point is to show that $\dim(\mathcal{M} \ominus z\mathcal{M}) \leq 1$.

For the time being, we suppose in addition that $\mathcal{M}$ contains a bounded function h such that $h(0) = 1$. We shall show how to relax this assumption at the end of the proof.

To prove that $\dim(\mathcal{M} \ominus z\mathcal{M}) \leq 1$, it is sufficient to show that, if $g \in \mathcal{M}$ and $g(0) = 0$, then $g \in z\mathcal{M}$. Fix $g \in \mathcal{M}$ with $g(0) = 0$. By Corollary 7.6.2, we can write $g = zf$, where $f \in \mathcal{D}_\mu$. We need to show that $f \in \mathcal{M}$. By Lemma 8.3.5 there exists a sequence (f_n) in $\mathcal{D}_\mu$ such that $|f_n| \leq |f|$ and $\|f_n - f\|_{\mathcal{D}_\mu} \to 0$. Consider the following decomposition:

$$f_n = f_n h + (zf)\Big(\frac{f_n}{f}\frac{1-h}{z}\Big).$$

The summands on the right-hand side belong to $\mathcal{D}_\mu$, because $f_n, h \in \mathcal{D}_\mu \cap H^\infty$, which is an algebra (see Theorem 1.6.6). Further, $f_n \in H^\infty$ and $h \in \mathcal{M}$, so by Theorem 8.3.2 we have $f_n h \in \mathcal{M}$. Also $zf \in \mathcal{M}$ and $(f_n/f)(1-h)/z \in H^\infty$, so a second application of Theorem 8.3.2 yields that their product too lies in $\mathcal{M}$. Putting all of this together, we deduce that $f_n \in \mathcal{M}$. Finally, as $f_n \to f$ in $\mathcal{D}_\mu$, we obtain $f \in \mathcal{M}$, as desired.

It remains to remove the supposition about the existence of h. By assumption $\mathcal{M} \neq \{0\}$, so there exists a greatest $k \geq 0$ such that $\mathcal{M} \subset z^k\mathcal{D}_\mu$. Then $\mathcal{M} = z^k\mathcal{N}$, for some subspace $\mathcal{N}$ of $\mathcal{D}_\mu$. It is routine to verify that $\mathcal{N} \in \text{Lat}(M_z, \mathcal{D}_\mu)$. Moreover $\mathcal{N}$ contains a function h with $h(0) = 1$. Using Theorem 8.3.4, we may further suppose that h is bounded. Then, by what we have proved above, $\dim(\mathcal{N} \ominus z\mathcal{N}) \leq 1$. By elementary linear algebra, it follows that

$$\dim(\mathcal{M} \ominus z\mathcal{M}) = \dim(z^k\mathcal{N} \ominus z^{k+1}\mathcal{N}) \leq \dim(\mathcal{N} \ominus z\mathcal{N}) \leq 1.$$

This completes the proof. □

With this result established, the door is now open for the application of Theorem 8.2.2. This is what it yields.

Theorem 8.3.7 *Let $\mathcal{M} \in \text{Lat}(M_z, \mathcal{D}_\mu)$, $\mathcal{M} \neq \{0\}$, and let $\phi \in (\mathcal{M} \ominus z\mathcal{M}) \setminus \{0\}$. Then $\mathcal{M} = \phi\mathcal{D}_{\mu_\phi}$, where $d\mu_\phi := |\phi^*|^2\, d\mu$.*

Proof We apply Theorem 8.2.2 with $(T, \mathcal{H}) = (M_z, \mathcal{M})$. There exists a finite measure ν on $\mathbb{T}$ and a unitary map $U : \mathcal{D}_\nu \to \mathcal{M}$ such that $U^{-1}TU = M_z$. Since $1 \in \mathcal{D}_\nu \ominus z\mathcal{D}_\nu$, it follows that $U(1)$ belongs to $\mathcal{M} \ominus z\mathcal{M}$, and is therefore a constant multiple of ϕ. Multiplying ϕ by a constant, we can suppose that $U(1) = \phi$. Then, for every polynomial p, we have $U(p) = p\phi$. As polynomials are dense in $\mathcal{D}_\nu$, it follows that $U(f) = f\phi$ for all $f \in \mathcal{D}_\nu$. In particular, $\mathcal{M} = \phi\mathcal{D}_\nu$. It remains to identify the measure ν.

As U is unitary, we have $\|p\phi\|_{\mathcal{D}_\mu} = \|p\|_{\mathcal{D}_\nu}$ for all polynomials p. Using (8.1) for both μ and ν, we deduce that $\int_{\mathbb{T}} |p|^2|\phi^*|^2\,d\mu = \int_{\mathbb{T}} |p|^2\,d\nu$ for all p. An argument similar to that in the uniqueness part of Theorem 8.2.2 now shows that $d\nu = |\phi^*|^2\,d\mu$, whence the result. □

Corollary 8.3.8 *Every $\mathcal{M} \in \operatorname{Lat}(M_z, \mathcal{D}_\mu)$ is cyclic.*

Proof The theorem shows that $\mathcal{M}$ is generated (as an invariant subspace) by any non-zero $\phi \in \mathcal{M} \ominus z\mathcal{M}$. □

At the beginning of the chapter, we set ourselves the goal to generalize Beurling's theorem on invariant subspaces from H^2 to $\mathcal{D}_\mu$. It is instructive to see how far we have come. Recall that H^2 is just the special case of $\mathcal{D}_\mu$ with $\mu = 0$. In this case, Theorem 8.3.7 tells us that every $\mathcal{M} \in \operatorname{Lat}(M_z, H^2)$ is of the form ϕH^2, where $\phi \in \mathcal{M} \ominus z\mathcal{M}$. What is still lacking is some information about the function ϕ. In the case of H^2, if $\mathcal{M} \in \operatorname{Lat}(M_z, H^2) \setminus \{0\}$, it turns out that $\mathcal{M} \ominus z\mathcal{M}$ is spanned by an inner function, which yields Beurling's result. However, this is no longer true in $\mathcal{D}_\mu$ or even in $\mathcal{D}$. Instead, we have the following partial replacement.

Theorem 8.3.9 *Let $\mathcal{M} \in \operatorname{Lat}(M_z, \mathcal{D}_\mu)$ and let $\phi \in \mathcal{M} \ominus z\mathcal{M}$. Then*

(i) *$\phi \in H^\infty$ and $\|\phi\|_{H^\infty} \le \|\phi\|_{\mathcal{D}_\mu}$,*
(ii) *ϕ is a multiplier of $\mathcal{D}_\mu$, and $\|\phi f\|_{\mathcal{D}_\mu} \le \|\phi\|_{\mathcal{D}_\mu}\|f\|_{\mathcal{D}_\mu}$ for all $f \in \mathcal{D}_\mu$.*

Proof (i) We may suppose that $\|\phi\|_{\mathcal{D}_\mu} = 1$. Let $\phi = \phi_i\phi_o$ be the inner-outer factorization of ϕ. Fix $t > 1$, and set $\psi := (\phi_o \wedge t)/\phi_o$, a bounded outer function. We shall now prove two inequalities for $\log|\psi(0)|$.

Just as in the proof of Lemma 8.3.5, we have

$$\mathcal{D}_\mu(\psi\phi) = \mathcal{D}_\mu(\phi_i(\phi_o \wedge t)) \le \mathcal{D}_\mu(\phi). \tag{8.7}$$

In particular $\psi\phi \in \mathcal{D}_\mu$, and so in fact $\psi\phi \in \mathcal{M}$, by Theorem 8.3.2. Now Theorem 8.3.7 tells us that $\mathcal{M} = \phi\mathcal{D}_{\mu_\phi}$. Thus $\psi \in \mathcal{D}_{\mu_\phi}$. Using Corollary 7.6.2, it follows that $\psi - \psi(0) \in z\mathcal{D}_{\mu_\phi}$, and hence $(\psi - \psi(0))\phi \in z\mathcal{M}$. Therefore $\langle \phi, (\psi - \psi(0))\phi\rangle_{\mathcal{D}_\mu} = 0$. Expanding this out, we get $\langle \phi, \psi\phi\rangle_{\mathcal{D}_\mu} = \overline{\psi(0)}$. By the Cauchy–Schwarz inequality, it follows that $|\psi(0)| \le \|\psi\phi\|_{\mathcal{D}_\mu}$. Using the elementary fact that $\log x \le (x^2 - 1)/2$, we get

$$\log|\psi(0)| \le \frac{1}{2}(\|\psi\phi\|^2_{\mathcal{D}_\mu} - 1) = \frac{1}{2}(\|\psi\phi\|^2_{\mathcal{D}_\mu} - \|\phi\|^2_{\mathcal{D}_\mu}) \le \frac{1}{2}(\|\psi\phi\|^2_{H^2} - \|\phi\|^2_{H^2}),$$

the last inequality coming from (8.7). Recalling the way that ψ was defined, we thus have

$$\log|\psi(0)| \leq -\frac{1}{2}\int_{|\phi^*|>t}(|\phi^*|^2 - t^2)\,dm, \tag{8.8}$$

where m denotes normalized Lebesgue measure on $\mathbb{T}$.

On the other hand, as ψ is an outer function, it satisfies

$$\log|1/\psi(0)| = \int_{\mathbb{T}} \log|1/\psi|\,dm \leq \frac{1}{2}\int_{\mathbb{T}}(|1/\psi|^2 - 1)\,dm,$$

where we have once again used the inequality $\log x \leq (x^2 - 1)/2$. It follows that

$$\log|1/\psi(0)| \leq \frac{1}{2}\int_{|\phi^*|>t}(|\phi^*|^2/t^2 - 1)\,dm. \tag{8.9}$$

If we combine the inequalities (8.8) and (8.9), then we obtain

$$\int_{|\phi^*|>t}(|\phi^*|^2 - t^2)\,dm \leq \frac{1}{t^2}\int_{|\phi^*|>t}(|\phi^*|^2 - t^2)\,dm.$$

This holds for each $t > 1$, which is only possible if $|\phi^*| \leq 1$ m-a.e. on $\mathbb{T}$. Thus ϕ is bounded and $\|\phi\|_{H^\infty} \leq 1$.

(ii) As already mentioned in (i), we have $\mathcal{M} = \phi\mathcal{D}_{\mu_\phi}$, where $d\mu_\phi = |\phi^*|^2\,d\mu$. As ϕ is bounded, $\mathcal{D}_{\mu_\phi} \supset \mathcal{D}_\mu$. Therefore $\phi\mathcal{D}_\mu \subset \phi\mathcal{D}_{\mu_\phi} = \mathcal{M} \subset \mathcal{D}_\mu$. It follows that ϕ is a multiplier of $\mathcal{D}_\mu$.

Assume again that $\|\phi\|_{\mathcal{D}_\mu} = 1$. By (the proof of) Theorem 8.3.7, if $f \in \mathcal{D}_\mu$, then $\|\phi f\|_{\mathcal{D}_\mu} = \|f\|_{\mathcal{D}_{\mu_\phi}}$. Also, since $\|\phi\|_{H^\infty} \leq \|\phi\|_{\mathcal{D}_\mu} \leq 1$, we have

$$\mathcal{D}_{\mu_\phi}(f) = \int_{\mathbb{T}} \mathcal{D}_\zeta(f)\,d\mu_\phi(\zeta) \leq \int_{\mathbb{T}} \mathcal{D}_\zeta(f)\,d\mu(\zeta) = \mathcal{D}_\mu(f).$$

Therefore $\|f\|_{\mathcal{D}_{\mu_\phi}} \leq \|f\|_{\mathcal{D}_\mu}$, and so, finally, $\|\phi f\|_{\mathcal{D}_\mu} \leq \|f\|_{\mathcal{D}_\mu}$. □

As a corollary, we obtain a generalization to $\mathcal{D}_\mu$ of Theorem 5.4.1.

Corollary 8.3.10 *Given $f \in \mathcal{D}_\mu$, there exists a multiplier ϕ of $\mathcal{D}_\mu$ such that f and ϕ have the same zero set in $\mathbb{D}$.*

Proof By Theorem 8.3.9, there is a multiplier ϕ of $\mathcal{D}_\mu$ such that $[f]_{\mathcal{D}_\mu} = [\phi]_{\mathcal{D}_\mu}$. Since f and ϕ generate the same closed invariant subspace of $\mathcal{D}_\mu$, they must have the same zero set in $\mathbb{D}$. □

Exercises 8.3

1. Let $\mathcal{M}, \mathcal{N} \in \operatorname{Lat}(M_z, \mathcal{D}_\mu)$ with $\mathcal{M}, \mathcal{N} \neq \{0\}$. Show that $\mathcal{M} \cap \mathcal{N} \neq \{0\}$. [Hint: Show first that $\mathcal{M} \cap H^\infty \neq \{0\}$ and $\mathcal{N} \cap H^\infty \neq \{0\}$.]

Notes on Chapter 8

§8.1

The basic properties of the shift operator on $\mathcal{D}_\mu$ were established by Richter in [96]. Lemma 8.1.5 is due to Richter and Sundberg [98]. The formula in Exercise 8.1.1 is taken from [115].

§8.2

The results in this section are due to Richter [95, 96]. The terminology *2-isometry* was introduced by Agler [3].

The problem of characterizing those operators unitarily equivalent to M_z on spaces of analytic functions has been studied intensively. We mention in particular a remarkable result of Cowen and Douglas [37] (see also the paper of Carlsson [30]) that characterizes those operators unitarily equivalent to the shift operator M_z on some Hilbert space of analytic functions on a domain in $\mathbb{C}$.

§8.3

The crucial Theorem 8.3.6 was first obtained in the case of $\mathcal{D}$ by Richter and Shields [97], and subsequently extended to $\mathcal{D}_\mu$ by Richter and Sundberg [99]. The remaining results in this section are due to Richter [96] and to Richter and Sundberg [98, 99]. The proof of Theorem 8.3.9 is taken from the thesis of Aleman [7].

Olin and Thomson [88] defined an operator T to be *cellular-indecomposable* if we have $\mathcal{M} \cap \mathcal{N} \neq \{0\}$ whenever $\mathcal{M}, \mathcal{N} \in \operatorname{Lat}(T, \mathcal{H}) \setminus \{0\}$. Thus Exercise 8.3.1 shows that $(M_z, \mathcal{D}_\mu)$ is cellular-indecomposable. This can be used to prove that $\dim(\mathcal{M} \ominus z\mathcal{M}) \leq 1$ for $\mathcal{M} \in \operatorname{Lat}(M_z, \mathcal{D}_\mu)$. The paper of Bourdon [22] contains some more general results along these lines.

9

Cyclicity

In the previous chapter we saw that every (closed) invariant subspace of $\mathcal{D}_\mu$ is cyclic, in other words, that it is generated by a single function in $\mathcal{D}_\mu$. In this chapter, we shall take the opposite point of view: starting with $f \in \mathcal{D}_\mu$, can we identify the invariant subspace that it generates? It is easy to see that g belongs to this invariant subspace if and only if there is a sequence of polynomials (p_n) such that $\|p_n f - g\|_{\mathcal{D}_\mu} \to 0$, but in practice it is often difficult to determine which g have this property. Therefore we seek other, more explicit descriptions of the invariant subspace generated by f.

In particular, we pose the question: which functions f generate the whole of $\mathcal{D}_\mu$? A complete answer to this is not known, even for the special case of the Dirichlet space $\mathcal{D}$. In fact the characterization of those functions cyclic for $\mathcal{D}$ is the subject of a conjecture involving the logarithmic capacity of boundary zero sets. We shall present some partial solutions to this conjecture.

9.1 Cyclicity in $\mathcal{D}_\mu$

We begin by formalizing the notions above. Let μ be a finite positive measure on $\mathbb{T}$, and let $\mathcal{D}_\mu$ be the associated harmonically weighted Dirichlet space. As usual, if μ is normalized Lebesgue measure, then $\mathcal{D}_\mu$ is just the classical Dirichlet space $\mathcal{D}$, so all the results of this section apply in particular to $\mathcal{D}$.

Definition 9.1.1 Given $S \subset \mathcal{D}_\mu$, the *invariant subspace generated by* S is the smallest (closed) invariant subspace of $\mathcal{D}_\mu$ that contains S, namely the intersection of all $\mathcal{M} \in \operatorname{Lat}(M_z, \mathcal{D}_\mu)$ such that $S \subset \mathcal{M}$. We denote it by $[S]_{\mathcal{D}_\mu}$. In the case when S is a singleton $\{f\}$, we write simply $[f]_{\mathcal{D}_\mu}$. Finally, we say f is *cyclic* for $\mathcal{D}_\mu$ if $[f]_{\mathcal{D}_\mu} = \mathcal{D}_\mu$.

We begin with a useful reformulation of Theorem 8.3.2.

Theorem 9.1.2 *Let $f, g \in \mathcal{D}_\mu$ with $|g| \le |f|$ on $\mathbb{D}$. Then $[g]_{\mathcal{D}_\mu} \subset [f]_{\mathcal{D}_\mu}$.*

Proof Let $\mathcal{M} := [f]_{\mathcal{D}_\mu}$ and $\phi := g/f$. Then $f \in \mathcal{M}$, $\phi \in H^\infty$ and $\phi f = g \in \mathcal{D}_\mu$, so by Theorem 8.3.2, $\phi f \in \mathcal{M}$. In other words $g \in [f]_{\mathcal{D}_\mu}$. It follows that $[g]_{\mathcal{D}_\mu} \subset [f]_{\mathcal{D}_\mu}$. □

The next result describes $[f]_{\mathcal{D}_\mu}$ in terms of the inner-outer factorization of f. Recall that, if $f \in \mathcal{D}_\mu$ and if $f = f_i f_o$ is its inner-outer factorization, then also $f_o \in \mathcal{D}_\mu$ and $\|f_o\|_{\mathcal{D}_\mu} \le \|f\|_{\mathcal{D}_\mu}$ (see Corollary 7.6.3).

Theorem 9.1.3 *Let $f \in \mathcal{D}_\mu$ and let $f = f_i f_o$ be its inner-outer factorization. Then*

$$[f]_{\mathcal{D}_\mu} = f_i[f_o]_{\mathcal{D}_\mu} \cap \mathcal{D}_\mu = [f_o]_{\mathcal{D}_\mu} \cap f_i H^2.$$

For the proof, we need two lemmas. The first of these formalizes an idea that we have already used a few times.

Lemma 9.1.4 *Let $\mathcal{M}$ be a closed subspace of $\mathcal{D}_\mu$. Suppose that there exists a sequence (f_n) in $\mathcal{M}$ such that $\sup_n \mathcal{D}_\mu(f_n) < \infty$ and $f_n \to f$ pointwise in $\mathbb{D}$. Then $f \in \mathcal{M}$.*

Proof Using the result of Exercise 7.2.2, we have $\|f_n - f_n(0)\|_{H^2}^2 \le 4\mathcal{D}_\mu(f_n)$, and consequently

$$\|f_n\|_{\mathcal{D}_\mu}^2 = |f_n(0)|^2 + \|f_n - f_n(0)\|_{H^2}^2 + \mathcal{D}_\mu(f_n) \le |f_n(0)|^2 + 5\mathcal{D}_\mu(f_n).$$

It follows that $\sup_n \|f_n\|_{\mathcal{D}_\mu} < \infty$. Therefore there exists a subsequence (f_{n_j}) weakly convergent in $\mathcal{D}_\mu$. As weak convergence implies pointwise convergence in $\mathbb{D}$, the weak limit must be f. In particular $f \in \mathcal{D}_\mu$. Finally, as $\mathcal{M}$ is weakly closed in $\mathcal{D}_\mu$, we have $f \in \mathcal{M}$. □

Lemma 9.1.5 *Let θ be an inner function and let h be an outer function. Suppose that both $\theta h, \theta h^2 \in \mathcal{D}_\mu$. Then $\theta h \in [\theta h^2]_{\mathcal{D}_\mu}$.*

Proof By Corollary 7.6.2 we have $h, h^2 \in \mathcal{D}_\mu$. Also, by Theorem 7.5.4 we have $\mathcal{D}_\mu(h \wedge h^2) \le 4\mathcal{D}_\mu(h)$. Let $n \ge 1$, replace h by nh throughout, and then divide by n^2 to obtain $\mathcal{D}_\mu(h \wedge nh^2) \le 4\mathcal{D}_\mu(h)$. A calculation similar to that in the proof of Corollary 7.6.4 then shows that $\mathcal{D}_\mu(\theta(h \wedge nh^2)) \le 4\mathcal{D}_\mu(\theta h)$. On the other hand, for all n we have $|\theta(h \wedge nh^2)| \le n|\theta h^2|$, so by Theorem 9.1.2 we get $\theta(h \wedge nh^2) \in [\theta h^2]_{\mathcal{D}_\mu}$. Now apply Lemma 9.1.4 to deduce that $\theta h \in [\theta h^2]_{\mathcal{D}_\mu}$. □

Proof of Theorem 9.1.3 We shall prove the three inclusions

$$[f]_{\mathcal{D}_\mu} \subset ([f_o]_{\mathcal{D}_\mu} \cap f_i H^2) \subset (f_i[f_o]_{\mathcal{D}_\mu} \cap \mathcal{D}_\mu) \subset [f]_{\mathcal{D}_\mu}.$$

The first inclusion is easy. Indeed, $[f]_{\mathcal{D}_\mu} \subset [f_o]_{\mathcal{D}_\mu}$ by Theorem 9.1.2, and obviously $[f]_{\mathcal{D}_\mu} \subset [f]_{H^2} = f_i H^2$. Hence $[f]_{\mathcal{D}_\mu} \subset [f_o]_{\mathcal{D}_\mu} \cap f_i H^2$.

For the second inclusion, we begin by remarking that, by Theorem 8.3.7, $[f_o]_{\mathcal{D}_\mu} = \phi \mathcal{D}_{\mu_\phi}$ for some $\phi \in \mathcal{D}_\mu$. Further, since f_o is outer, ϕ must be outer as well. Thus, if $g \subset [f_o]_{\mathcal{D}_\mu} \cap f_i H^2$, then $g = \phi h$, where $h \in \mathcal{D}_{\mu_\phi}$ and $h \in f_i H^2$. By Corollary 7.6.2, it follows that $h = f_i k$, where $k \in \mathcal{D}_{\mu_\phi}$. Therefore $g = f_i \phi k \in f_i \phi \mathcal{D}_{\mu_\phi} = f_i [f_o]_{\mathcal{D}_\mu}$. Clearly also $g \in \mathcal{D}_\mu$. Thus $[f_o]_{\mathcal{D}_\mu} \cap f_i H^2 \subset f_i [f_o]_{\mathcal{D}_\mu} \cap \mathcal{D}_\mu$.

For the third inclusion, we note that $f_i [f_o]_{\mathcal{D}_\mu} \cap \mathcal{D}_\mu$ is a closed invariant subspace of $\mathcal{D}_\mu$, so it has the form $[\psi]_{\mathcal{D}_\mu}$ for some $\psi \in \mathcal{D}_\mu$. By Theorem 8.3.9, we may further suppose that ψ is a bounded multiplier of $\mathcal{D}_\mu$. Let $\psi = \psi_i \psi_o$ be the inner-outer factorization of ψ. Since both $f \in [\psi]_{\mathcal{D}_\mu} \subset [\psi]_{H^2} = \psi_i H^2$ and $\psi \in f_i [f_o]_{\mathcal{D}_\mu} \subset f_i H^2$, we must have $\psi_i = f_i$. It follows that $\psi_o \in [f_o]_{\mathcal{D}_\mu}$. Thus there exist polynomials (p_n) such that $p_n f_o \to \psi_o$ in $\mathcal{D}_\mu$. As ψ is a multiplier, $p_n f_o \psi \to \psi_o \psi$ in $\mathcal{D}_\mu$. Also, again using the fact that $\psi_i = f_i$, we have $p_n f_o \psi = p_n \psi_o f \in [f]_{\mathcal{D}_\mu}$. Consequently $\psi_o \psi \in [f]_{\mathcal{D}_\mu}$. By Lemma 9.1.5, $\psi \in [\psi \psi_o]_{\mathcal{D}_\mu}$. Therefore $\psi \in [f]_{\mathcal{D}_\mu}$. As ψ generates $f_i [f_o]_{\mathcal{D}_\mu} \cap \mathcal{D}_\mu$, we conclude that $f_i [f_o]_{\mathcal{D}_\mu} \cap \mathcal{D}_\mu \subset [f]_{\mathcal{D}_\mu}$. □

Theorem 9.1.3 effectively reduces the general problem of identifying $[f]_{\mathcal{D}_\mu}$ to the special case where f is an outer function. In the Hardy space, every outer function f satisfies $[f]_{H^2} = H^2$. This is no longer true in $\mathcal{D}_\mu$ (a simple example is given in Exercise 9.1.1) or even in $\mathcal{D}$ (more on this in the next section).

In the rest of the section, we gather together a few general results about invariant subspaces of $\mathcal{D}_\mu$ generated by outer functions. The first of these allows us in many cases to restrict our attention to bounded functions.

Theorem 9.1.6 *Let f be an outer function in $\mathcal{D}_\mu$. Then $[f \wedge 1]_{\mathcal{D}_\mu} = [f]_{\mathcal{D}_\mu}$.*

Proof The inclusion $[f \wedge 1]_{\mathcal{D}_\mu} \subset [f]_{\mathcal{D}_\mu}$ is immediate from Theorem 9.1.2. Also, since $|f \wedge n| \le n|f \wedge 1|$, the same theorem shows that $f \wedge n \in [f \wedge 1]_{\mathcal{D}_\mu}$ for each n, and, using Theorem 7.5.2, we have $\mathcal{D}_\mu(f \wedge n) \le \mathcal{D}_\mu(f)$ for all n. By Lemma 9.1.4, $f \in [f \wedge 1]_{\mathcal{D}_\mu}$. □

Theorem 9.1.7 *Let f be an outer function, let $\alpha > 0$, and suppose $f, f^\alpha \in \mathcal{D}_\mu$. Then $[f^\alpha]_{\mathcal{D}_\mu} = [f]_{\mathcal{D}_\mu}$.*

Proof We first prove the result under the additional assumption that f is bounded. Then $f^n \in \mathcal{D}_\mu$ for all n, because $\mathcal{D}_\mu \cap H^\infty$ is an algebra. If $\alpha \ge 1$ and $n \ge \alpha$, then Theorem 9.1.2 shows that $[f^n]_{\mathcal{D}_\mu} \subset [f^\alpha]_{\mathcal{D}_\mu} \subset [f]_{\mathcal{D}_\mu}$. Also, repeated application of Lemma 9.1.5 shows that $[f]_{\mathcal{D}_\mu} \subset [f^2]_{\mathcal{D}_\mu} \subset \cdots \subset [f^{2^k}]_{\mathcal{D}_\mu}$ for all k. Combining these inclusions we obtain $[f]_{\mathcal{D}_\mu} = [f^\alpha]_{\mathcal{D}_\mu}$ for all $\alpha \ge 1$. The case $\alpha < 1$ follows by switching the roles of f and f^α.

Finally, if f is unbounded, then we use Theorem 9.1.6 twice over, as follows:

$$[f]_{\mathcal{D}_\mu} = [f \wedge 1]_{\mathcal{D}_\mu} = [(f \wedge 1)^\alpha]_{\mathcal{D}_\mu} = [f^\alpha \wedge 1]_{\mathcal{D}_\mu} = [f^\alpha]_{\mathcal{D}_\mu}. \qquad \square$$

Theorem 9.1.8 *Let f, g be outer functions in $\mathcal{D}_\mu$. Then $[f \vee g]_{\mathcal{D}_\mu} = [f, g]_{\mathcal{D}_\mu}$.*

Proof By Theorem 9.1.2, we have $f, g \in [f \vee g]_{\mathcal{D}_\mu}$ and so $[f, g]_{\mathcal{D}_\mu} \subset [f \vee g]_{\mathcal{D}_\mu}$. For the reverse inclusion, we use Theorem 8.3.7 to write $[f, g]_{\mathcal{D}_\mu} = \phi \mathcal{D}_{\mu_\phi}$, where $\phi \in \mathcal{D}_\mu$. Then $f = \phi f_1$ and $g = \phi g_1$, where $f_1, g_1 \in \mathcal{D}_{\mu_\phi}$. As f, g are outer, so are ϕ, f_1, g_1. By Theorem 7.5.2, we have $f_1 \vee g_1 \in \mathcal{D}_{\mu_\phi}$. Hence $f \vee g = \phi(f_1 \vee g_1) \in \phi \mathcal{D}_{\mu_\phi} = [f, g]_{\mathcal{D}_\mu}$, and so finally $[f \vee g]_{\mathcal{D}_\mu} \subset [f, g]_{\mathcal{D}_\mu}$. $\square$

Theorem 9.1.9 *Let f, g be outer functions, and suppose that $f, g, fg \in \mathcal{D}_\mu$. Then $[fg]_{\mathcal{D}_\mu} = [f]_{\mathcal{D}_\mu} \cap [g]_{\mathcal{D}_\mu}$.*

Proof We begin by showing that $[f]_{\mathcal{D}_\mu} \cap [g]_{\mathcal{D}_\mu} \subset [fg]_{\mathcal{D}_\mu}$.

As $[f]_{\mathcal{D}_\mu} \cap [g]_{\mathcal{D}_\mu}$ is an invariant subspace, it must be of the form $[\phi]_{\mathcal{D}_\mu}$, where ϕ is a (bounded) multiplier of $\mathcal{D}_\mu$. We may suppose that $\|\phi\|_\infty < 1$. Also, since $[f]_{\mathcal{D}_\mu} \cap [g]_{\mathcal{D}_\mu}$ contains at least one outer function, for example $(f \wedge 1)(g \wedge 1)$, it follows that ϕ must be outer. Consequently $[\phi^2]_{\mathcal{D}_\mu} = [\phi]_{\mathcal{D}_\mu}$, by Theorem 9.1.7. Thus, if we can show that $\phi^2 \in [fg]_{\mathcal{D}_\mu}$, then we will have proved the inclusion $[f]_{\mathcal{D}_\mu} \cap [g]_{\mathcal{D}_\mu} \subset [fg]_{\mathcal{D}_\mu}$.

Since $\phi \in [f]_{\mathcal{D}_\mu} \cap [g]_{\mathcal{D}_\mu}$, there are sequences of polynomials (p_n) and (q_n) such that $p_n f \to \phi$ and $q_n g \to \phi$ in $\mathcal{D}_\mu$. Writing $p_n f = (p_n f)_i (p_n f)_o$ for the inner-outer factorization, we set $f_n := (p_n f)_i((p_n f)_o \wedge 1)$. By Corollary 7.6.4 we have $\mathcal{D}_\mu(f_n) \le \mathcal{D}_\mu(p_n f)$ for all n. As $p_n f \to \phi$ in $\mathcal{D}_\mu$, it follows that $\sup_n \mathcal{D}_\mu(f_n) < \infty$. Likewise, defining $g_n := (g_n g)_i((q_n g)_o \wedge 1)$, we obtain $\sup_n \mathcal{D}_\mu(g_n) < \infty$. Since $\|f_n\|_\infty \le 1$ and $\|g_n\|_\infty \le 1$ for all n, we deduce that $\sup_n \mathcal{D}_\mu(f_n g_n) < \infty$. For each n, there exists a constant C_n such that $|f_n g_n| \le C_n |fg|$, so $f_n g_n \in [fg]_{\mathcal{D}_\mu}$. Clearly $f_n g_n$ converges pointwise in $\mathbb{D}$ to ϕ^2. Applying Lemma 9.1.4, we get $\phi^2 \in [fg]_{\mathcal{D}_\mu}$, as desired.

Now we turn to the proof of the reverse inclusion $[fg]_{\mathcal{D}_\mu} \subset [f]_{\mathcal{D}_\mu} \cap [g]_{\mathcal{D}_\mu}$. This follows directly from Theorem 9.1.2 if f, g are both bounded, so the proof boils down to relaxing the boundedness of f, g. We do this in two stages.

Suppose first that f is unbounded, but g is still bounded. Then, for each n we have $(f \wedge n)g \in [f]_{\mathcal{D}_\mu} \cap [g]_{\mathcal{D}_\mu}$. We claim that $\sup_n \mathcal{D}_\mu((f \wedge n)g) < \infty$. If so, then Lemma 9.1.4 shows that $fg \in [f]_{\mathcal{D}_\mu} \cap [g]_{\mathcal{D}_\mu}$. To justify the claim, we use Lemma 8.1.5. According to that lemma, for μ-almost every $\zeta \in \mathbb{T}$, we have

$$\begin{aligned}
\mathcal{D}_\zeta((f \wedge n)g) &\le 2\|g\|_{H^\infty}^2 \mathcal{D}_\zeta(f \wedge n) + 2\mathcal{D}_\zeta(g)|(f \wedge n)^*(\zeta)|^2 \\
&\le 2\|g\|_{H^\infty}^2 \mathcal{D}_\zeta(f) + 2\mathcal{D}_\zeta(g)|f^*(\zeta)|^2 \\
&\le 6\|g\|_{H^\infty}^2 \mathcal{D}_\zeta(f) + 4\mathcal{D}_\zeta(fg).
\end{aligned}$$

Integrating with respect to $d\mu(\zeta)$ gives

$$\mathcal{D}_\mu((f \wedge n)g) \le 6\|g\|_{H^\infty}^2 \mathcal{D}_\mu(f) + 4\mathcal{D}_\mu(fg).$$

In particular, $\sup_n \mathcal{D}_\mu((f \wedge n)g) < \infty$, justifying the claim.

Now suppose both f and g are unbounded. Note that $f(g \wedge 1) = fg \wedge f \in \mathcal{D}_\mu$. Hence, by what we have just proved, $[f(g \wedge 1)]_{\mathcal{D}_\mu} \subset [g]_{\mathcal{D}_\mu}$. Clearly we have $|fg \wedge 1| \le |f(g \wedge 1) \vee g|$. Hence, from Theorems 9.1.6, 9.1.2 and 9.1.8, we get

$$[fg]_{\mathcal{D}_\mu} = [fg \wedge 1]_{\mathcal{D}_\mu} \subset [f(g \wedge 1) \vee g]_{\mathcal{D}_\mu} = [f(g \wedge 1), g]_{\mathcal{D}_\mu} \subset [g]_{\mathcal{D}_\mu}.$$

By symmetry $[fg]_{\mathcal{D}_\mu} \in [f]_{\mathcal{D}_\mu}$ as well, whence the result. □

Corollary 9.1.10 *If $f \in \mathcal{D}_\mu$ is invertible in $\mathcal{D}_\mu$, then it is cyclic for $\mathcal{D}_\mu$.*

Proof To say that f is invertible means that $1/f \in \mathcal{D}_\mu$. This forces both f and $1/f$ to be outer functions. By Theorem 9.1.9, we then have $[f]_{\mathcal{D}_\mu} \cap [1/f]_{\mathcal{D}_\mu} = [1]_{\mathcal{D}_\mu} = \mathcal{D}_\mu$. In particular $[f]_{\mathcal{D}_\mu} = \mathcal{D}_\mu$. □

The final result of the section is the companion to Theorem 9.1.8 and also generalizes Theorem 9.1.6.

Theorem 9.1.11 *Let f and g be outer functions in $\mathcal{D}_\mu$. Then $[f \wedge g]_{\mathcal{D}_\mu} = [f]_{\mathcal{D}_\mu} \cap [g]_{\mathcal{D}_\mu}$.*

Proof By Theorem 9.1.2 we have $[f \wedge g]_{\mathcal{D}_\mu} \subset [f]_{\mathcal{D}_\mu} \cap [g]_{\mathcal{D}_\mu}$. The reverse inclusion follows from

$$[f]_{\mathcal{D}_\mu} \cap [g]_{\mathcal{D}_\mu} = [f \wedge 1]_{\mathcal{D}_\mu} \cap [g \wedge 1]_{\mathcal{D}_\mu} = [(f \wedge 1)(g \wedge 1)]_{\mathcal{D}_\mu} \subset [f \wedge g]_{\mathcal{D}_\mu},$$

where we used successively Theorems 9.1.6, 9.1.9 and 9.1.2. □

Exercises 9.1

1. Let $\mu = \delta_1$, the Dirac mass at $\{1\}$. Let $\mathcal{M} := \{f \in \mathcal{D}_\mu : f^*(1) = 0\}$. Show that $\mathcal{M}$ is a closed invariant subspace of $(M_z, \mathcal{D}_\mu)$. Deduce that $[z-1]_{\mathcal{D}_\mu} \ne \mathcal{D}_\mu$, even though $z - 1$ is an outer function.
2. Let $f \in \mathcal{D}_\mu$, and suppose also that $\log f \in \mathcal{D}_\mu$.
   (i) Show that f is an outer function.
   (ii) Show that, for all $\alpha \in (0, 1)$, we have
   $$\mathcal{D}_\mu(f^\alpha) \le \alpha^2(\mathcal{D}_\mu(f) + \mathcal{D}_\mu(\log f)).$$
   (iii) Deduce that $f^\alpha \to 1$ in $\mathcal{D}_\mu$ as $\alpha \to 0^+$.
   (iv) Deduce that f is cyclic for $\mathcal{D}_\mu$.
3. Give an example to show that a function can be cyclic for $\mathcal{D}_\mu$ without being invertible in $\mathcal{D}_\mu$.

4. Let $f, g \in \mathcal{D}_\mu$, and suppose also that $fg \in \mathcal{D}_\mu$.
   (i) Show that, if at least one of f, g is outer, then $[fg]_{\mathcal{D}_\mu} = [f]_{\mathcal{D}_\mu} \cap [g]_{\mathcal{D}_\mu}$.
   (ii) Give an example to show that, if neither f nor g is outer, then we may have $[fg]_{\mathcal{D}_\mu} \neq [f]_{\mathcal{D}_\mu} \cap [g]_{\mathcal{D}_\mu}$.

9.2 Cyclicity in $\mathcal{D}$ and boundary zero sets

The results of the previous section relate cyclic subspaces of $\mathcal{D}_\mu$ to others but, with the exception of Corollary 9.1.10, they do not explicitly identify $[f]_{\mathcal{D}_\mu}$. This is because the answer depends on the measure μ. To proceed further, we need to know what μ is. In this section, we consider the important case where μ is Lebesgue measure, and so $\mathcal{D}_\mu = \mathcal{D}$, the classical Dirichlet space.

We begin with a basic example.

Theorem 9.2.1 *Let p be a polynomial. Then*

$$[p]_{\mathcal{D}} = \{f \in \mathcal{D} : f(w_j) = 0,\ j = 1, \dots, m\},$$

where $w_1, \dots, w_m$ are the zeros of p inside $\mathbb{D}$, counted according to multiplicity. In particular, if p has no zeros in $\mathbb{D}$, then it is cyclic for $\mathcal{D}$.

Proof Let $p = p_i p_o$ be the inner-outer factorization of p. Here the inner factor p_i is the finite Blaschke product with zeros at $w_1, \dots, w_m$, and the outer factor p_o is a polynomial all of whose zeros lie in $\mathbb{C} \setminus \mathbb{D}$. By Theorem 9.1.3,

$$[p]_{\mathcal{D}} = [p_o]_{\mathcal{D}} \cap p_i H^2 = [p_o]_{\mathcal{D}} \cap \{f \in H^2 : f(w_j) = 0,\ j = 1, \dots, m\}.$$

It therefore suffices to show that $[p_o]_{\mathcal{D}} = \mathcal{D}$.

Let us write $p_o(z) = c(z - a_1) \dots (z - a_n)$, where $a_1, \dots, a_n \in \mathbb{C} \setminus \mathbb{D}$, and $c \in \mathbb{C}$ with $c \neq 0$. By Theorem 9.1.9, we have

$$[p_o]_{\mathcal{D}} = [c]_{\mathcal{D}} \cap [z - a_1]_{\mathcal{D}} \cap \cdots \cap [z - a_n]_{\mathcal{D}}.$$

Clearly $[c]_{\mathcal{D}} = \mathcal{D}$. So it suffices to show that $[z - a]_{\mathcal{D}} = \mathcal{D}$ for all $a \in \mathbb{C} \setminus \mathbb{D}$.

Fix $a \in \mathbb{C} \setminus \mathbb{D}$. Let $g \in \mathcal{D} \ominus [z - a]_{\mathcal{D}}$, say $g(z) = \sum_{n \geq 0} b_n z^n$. Then we have $\langle g, z^{n-1}(z - a) \rangle_{\mathcal{D}} = 0$ for all $n \geq 1$. This translates to $(n + 1)b_n - nab_{n-1} = 0$ for all $n \geq 1$, and consequently $(n + 1)b_n = a^n b_0$ for all $n \geq 0$. Thus

$$\|g\|_{\mathcal{D}}^2 = \sum_{n \geq 0} (n + 1)|b_n|^2 = \sum_{n \geq 0} |a|^{2n} |b_0|^2 / (n + 1),$$

and this can only be finite if $b_0 = 0$, whence $g = 0$. Therefore $[z - a]_{\mathcal{D}} = \mathcal{D}$. □

One way to interpret this result is that a polynomial is cyclic for $\mathcal{D}$ if and only if it is an outer function. One might guess that the same is true for general functions in $\mathcal{D}$ (as is the case in H^2) but, as we shall see, this turns out to be false. The reason is that zero sets on the boundary also have a role to play. To discuss this further, it is convenient to introduce some notation.

Recall that q.e. means quasi-everywhere, namely outside a set of outer logarithmic capacity zero.

Definition 9.2.2 Given an arbitrary subset E of $\mathbb{T}$, we write

$$\mathcal{D}_E := \{f \in \mathcal{D} : f^* = 0 \text{ q.e. on } E\}.$$

Clearly $\mathcal{D}_E = \mathcal{D}$ if and only if $c^*(E) = 0$. Also $\mathcal{D}_E = \{0\}$ whenever $|E| > 0$. We shall return to study the intermediate cases later.

The following theorem provides the connection between boundary zero sets and invariant subspaces.

Theorem 9.2.3 *For every subset $E \subset \mathbb{T}$, we have $\mathcal{D}_E \in \operatorname{Lat}(M_z, \mathcal{D})$.*

Proof It is clear that $\mathcal{D}_E$ is a subspace of $\mathcal{D}$ and that it is M_z-invariant. The point at issue is whether it is closed in $\mathcal{D}$.

Let (f_n) be a sequence in $\mathcal{D}_E$, and suppose that $f_n \to f$ in $\mathcal{D}$. By the weak-type inequality for capacity Theorem 3.3.1, for each $t > 0$ we have

$$c^*(E \cap \{|f^*| > t\}) \le c^*(|f^* - f_n^*| > t) \le A\|f - f_n\|_{\mathcal{D}}^2/t^2,$$

where A is an absolute constant. Let $n \to \infty$ to get $c^*(E \cap \{|f^*| > t\}) = 0$. As this holds for all $t > 0$, it follows that $f^* = 0$ q.e. on E, in other words, $f \in \mathcal{D}_E$. Thus $\mathcal{D}_E$ is closed in $\mathcal{D}$, as claimed. □

Corollary 9.2.4 *If $f \in \mathcal{D}$, then $[f]_{\mathcal{D}} \subset \mathcal{D}_E$, where $E := \{\zeta \in \mathbb{T} : f^*(\zeta) = 0\}$.*

Proof Clearly $f \in \mathcal{D}_E$, and by the theorem $\mathcal{D}_E \in \operatorname{Lat}(M_z, \mathcal{D})$. □

Corollary 9.2.5 *If f is cyclic for $\mathcal{D}$, then f is outer and $c^*(f^* = 0) = 0$.*

Proof The first conclusion follows from (the easy part of) Theorem 9.1.3, and the second one follows from Corollary 9.2.4. □

As a simple application of these ideas, we can prove an analogue of Theorem 5.4.1 for boundary zero sets.

Theorem 9.2.6 *Given $f \in \mathcal{D}$, there exists a multiplier ϕ of $\mathcal{D}$ such that the boundary zero sets $\{f^* = 0\}$ and $\{\phi^* = 0\}$ differ by a set of outer logarithmic capacity zero.*

Proof By Theorem 8.3.9 there exists $\phi \in \mathcal{M}(\mathcal{D})$ such that $[f]_{\mathcal{D}} = [\phi]_{\mathcal{D}}$. Corollary 9.2.4 then tells us that $f^* = 0$ q.e. on $\{\phi^* = 0\}$ and $\phi^* = 0$ q.e. on $\{f^* = 0\}$. In other words, the symmetric difference of $\{f^* = 0\}$ and $\{\phi^* = 0\}$ is of outer capacity zero. □

In §4.3 and §4.4 we gave some constructions of boundary zero sets for $\mathcal{D}$. What do these constructions tell us about cyclicity? We shall derive two consequences. The first shows that the conclusion in Corollary 9.2.5 cannot be improved (at least for $f \in \mathcal{D} \cap A(\mathbb{D})$), and the second shows that no part of the conclusion is redundant.

Theorem 9.2.7 *Let E be a closed subset of $\mathbb{T}$ such that $c(E) = 0$. Then there exists $f \in \mathcal{D} \cap A(\mathbb{D})$ such that $f = 0$ on E and f is cyclic for $\mathcal{D}$.*

Proof The existence of f, without the cyclicity statement, was established in Theorem 4.3.2. using the construction in Theorem 3.4.1. We shall now prove that f, as constructed there, is cyclic.

The function f was built as $f = \exp(-\sum_{j\ge1} F_j)$, where $F_j \in \mathcal{D} \cap A(\mathbb{D})$ with $\operatorname{Re} F_j \ge 0$ and $\sum_j \|F_j\|_{\mathcal{D}} < \infty$. Let $f_n := \exp(-\sum_{j>n} F_j)$. Then we have $\mathcal{D}(f_n) \le \mathcal{D}(\sum_{j>n} F_j) \le (\sum_{j>n} \|F_j\|_{\mathcal{D}})^2 \to 0$, and $f_n \to 1$ boundedly on $\mathbb{D}$, so $\|f_n - 1\|_{\mathcal{D}} \to 0$. Also, for each n, we have $f_n/f = \exp(\sum_{j=1}^n F_j)$, which is a bounded function, so $f_n \in [f]_{\mathcal{D}}$. Letting $n \to \infty$, we deduce that $1 \in [f]_{\mathcal{D}}$. □

Theorem 9.2.8 *There exists an outer function $f \in \mathcal{D} \cap A(\mathbb{D})$ which is not cyclic for $\mathcal{D}$.*

Proof Let E be the circular Cantor middle-third set. Then E is a Carleson set, so by Theorem 4.4.3, there exists an outer function $f \in \mathcal{D} \cap A(\mathbb{D})$ whose zero set is precisely E. Also, $c(E) > 0$, so by Corollary 9.2.5 it follows that f is not cyclic for $\mathcal{D}$. □

Exercises 9.2

1. For $n \ge 1$, set $L_n := \sum_{k=1}^n 1/k$ and

$$p_n(z) := \sum_{k=1}^{n} \Big(1 - \frac{L_k}{L_n}\Big) z^{k-1}.$$

Show by direct computation that

$$\|1 - (1 - z)p_n(z)\|_{\mathcal{D}}^2 = 1/L_n.$$

Deduce that $(1 - z)$ is cyclic for $\mathcal{D}$.

2. Let E be a countable subset of $\mathbb{T}$ (not necessarily closed). Show that there exists a cyclic function $f \in \mathcal{D}$ such that $\lim_{r\to 1} f(r\zeta) = 0$ for all $\zeta \in E$. [Hint: Use Theorem 4.3.1.]
3. Let $f \in \mathcal{D}$ be a univalent function such that $f(z) \neq 0$ for all $z \in \mathbb{D}$.

   (i) Prove that f is an outer function. [Hint: By Theorem A.1.10 in Appendix A, we have $1/f \in H^p$ for all $p < 1/2$.]
   (ii) Use the change of variable $w := f(z)$ to show that, for $0 < \alpha < 1$,

$$\mathcal{D}(f^\alpha) \leq \alpha + \alpha^2 \mathcal{D}(f).$$

   (iii) Deduce that $f^\alpha \to 1$ in $\mathcal{D}$ as $\alpha \to 0^+$.
   (iv) Deduce that f is cyclic for $\mathcal{D}$.
   (v) Deduce that $c^*(f^* = 0) = 0$.

9.3 The Brown–Shields conjecture

In the previous section, we saw that there are two obstacles to cyclicity in $\mathcal{D}$: inner factors and zero sets on the boundary. Are there any others? We formulate this question more precisely, as follows.

Problem 9.3.1 *Let $f \in \mathcal{D}$ be outer and let $E := \{f^* = 0\}$. Is $[f]_\mathcal{D} = \mathcal{D}_E$?*

The inclusion $[f]_\mathcal{D} \subset \mathcal{D}_E$ always holds, by Corollary 9.2.4. The issue is whether $[f]_\mathcal{D} \supset \mathcal{D}_E$.

Problem 9.3.1 seems to have been raised first by Brown and Shields in [24]. They also conjectured that the following special case should always hold.

Conjecture 9.3.2 (Brown–Shields) *If $f \in \mathcal{D}$ is outer and $c^*(f^* = 0) = 0$, then f is cyclic for $\mathcal{D}$.*

In other words, the two necessary conditions for cyclicity in Corollary 9.2.5 are between them also sufficient. This conjecture (and *a fortiori* Problem 9.3.1) are still open. In this section and the next, we present some partial solutions, starting with the following result.

Theorem 9.3.3 *Let $f \in \mathcal{D}$ be an outer function, and define*

$$E := \{\zeta \in \mathbb{T} : \liminf_{z\to\zeta} |f(z)| = 0\}. \tag{9.1}$$

If $g \in \mathcal{D}$ and $|g(z)| \leq C \operatorname{dist}(z, E)$ on $\mathbb{D}$, then $g \in [f]_\mathcal{D}$.

Theorem 9.3.3 falls short of being a true solution to Problem 9.3.1 for two reasons.

Firstly, the set E is defined differently. In the conjecture it is the zero set of f^*, and in Theorem 9.3.3 it is the zero set of $\liminf |f|$, which may in principle be a larger set. Of course, the two sets agree if f (or even just $|f|$) is continuous up to the boundary. However, there are examples of outer functions in $\mathcal{D}$ where the two sets disagree, and the set in (9.1) may even be the whole unit circle (see Exercise 9.3.2).

Secondly, in the theorem g is assumed to satisfy the Lipschitz-type estimate $|g(z)| \le C \operatorname{dist}(z, E)$, rather than the much weaker condition $g^* = 0$ q.e. on E defining membership of $\mathcal{D}_E$. The Lipschitz condition forces E to be a Carleson set (see Exercise 9.3.3), so it is quite restrictive.

To prove the theorem, we require the following 'fusion lemma'. We write d for arclength distance on $\mathbb{T}$, and $|\cdot|$ for Lebesgue measure on $\mathbb{T}$.

Lemma 9.3.4 *Let $f_1, f_2 \in \mathcal{D}$ be outer functions such that $|f_j^*(\zeta)| \le C d(\zeta, E)$, where E is a closed subset of $\mathbb{T}$. Let f be an outer function such that, on each component of $\mathbb{T} \setminus E$, either $|f^*| = |f_1^*|$ a.e. or $|f^*| = |f_2^*|$ a.e. Then $f \in \mathcal{D}$, and*

$$\mathcal{D}(f) \le \mathcal{D}(f_1) + \mathcal{D}(f_2) + \frac{C^2\pi^2}{2} \log\Big(\frac{C^2\pi^2}{|f_1(0)f_2(0)|}\Big).$$

The precise value of the bound is unimportant. What matters is that it is independent of E.

Proof By scaling, we can suppose that $C = 1/\pi$. This implies that $|f_j| \le 1$. Also, $|E| = 0$ and $f \in H^2$. Thus, by Carleson's formula (Theorem 7.4.1),

$$\begin{aligned}\mathcal{D}(f) &= \iint_{\mathbb{T}\times\mathbb{T}} \frac{1}{4\pi^2} \frac{(|f^*(\lambda)|^2 - |f^*(\zeta)|^2)(\log|f^*(\lambda)| - \log|f^*(\zeta)|)}{|\lambda - \zeta|^2} |d\lambda|\,|d\zeta| \\ &= \iint_{U_1\times U_1} + \iint_{U_2\times U_2} + 2\iint_{U_1\times U_2},\end{aligned}$$

where U_j is the union of the components of $\mathbb{T} \setminus E$ on which $|f^*| = |f_j^*|$ a.e. By Carleson's formula again, the first two terms on the right-hand side are bounded above by $\mathcal{D}(f_1)$ and $\mathcal{D}(f_2)$ respectively. It remains to estimate the third term, namely

$$\frac{1}{2\pi^2} \int_{U_1}\int_{U_2} \frac{(|f_1^*(\lambda)|^2 - |f_2^*(\zeta)|^2)(\log|f_1^*(\lambda)| - \log|f_2^*(\zeta)|)}{|\lambda - \zeta|^2} |d\lambda|\,|d\zeta|. \tag{9.2}$$

If $\lambda \in U_1$ and $\zeta \in U_2$, then there must be a point of E between them, so $d(\lambda, \zeta) \ge d(\lambda, E) + d(\zeta, E)$, and consequently

$$\Big||f_1^*(\lambda)|^2 - |f_2^*(\zeta)|^2\Big| \le \pi^{-2} d(\lambda, E)^2 + \pi^{-2} d(\zeta, E)^2 \le \pi^{-2} d(\lambda, \zeta)^2 \le \frac{1}{4}|\lambda - \zeta|^2.$$

Therefore the expression in (9.2) is bounded above by

$$\begin{aligned}\frac{1}{8\pi^2}&\iint_{U_1\times U_2}\Big|\log|f_1^*(\lambda)|-\log|f_2^*(\zeta)|\Big|\,|d\lambda|\,|d\zeta|\\ &\le\frac{1}{8\pi^2}\iint_{\mathbb{T}\times\mathbb{T}}\Big(\log\frac{1}{|f_1^*(\lambda)|}+\log\frac{1}{|f_2^*(\zeta)|}\Big)\,|d\lambda|\,|d\zeta|\\ &=\frac{1}{2}\log\frac{1}{|f_1(0)|}+\frac{1}{2}\log\frac{1}{|f_2(0)|},\end{aligned}$$

where we have used the facts that $|f_j^*|\le 1$ and f_j is outer ($j=1,2$). This gives the required estimate. □

Proof of Theorem 9.3.3 We can assume that f is bounded. Otherwise replace f by $f\wedge 1$ and use Theorem 9.1.6. This does not change the set E (see Exercise 9.3.1). Note that, as g is also bounded, it follows that $fg\in\mathcal{D}$.

Let $(I_n)_{n\ge1}$ be the connected components of $\mathbb{T}\setminus E$. For each n, let g_n be the outer function such that

$$|g_n^*|=\begin{cases}|g^*|, & \text{a.e. on } \cup_{j\le n} I_j,\\ |f^*g^*|, & \text{a.e. on } \cup_{j>n} I_j.\end{cases}$$

Clearly $g_n\to g_o$ pointwise on $\mathbb{D}$, where g_o is the outer factor of g. Also, by Lemma 9.3.4, we have $\sup_n\mathcal{D}(g_n)<\infty$. We claim that $g_n\in[f]_{\mathcal{D}}$ for all n. If so, then by Lemma 9.1.4 we obtain $g_o\in[f]_{\mathcal{D}}$, and hence also $g\in[f]_{\mathcal{D}}$, thereby proving the theorem.

It remains to establish the claim that $g_n\in[f]_{\mathcal{D}}$. Fix n. Let $(J_k)_{k\ge1}$ be an increasing sequence of compact sets, each a union of n arcs, such that $\cup_{k\ge1}J_k=\cup_{j=1}^n I_j$. For each k, let p_k be the monic polynomial of degree $2n$ with zeros at the endpoints of J_k, and let h_k be the outer function such that

$$|h_k^*|=\begin{cases}|p_kg^*|, & \text{a.e. on } J_k,\\ |p_kf^*g^*|, & \text{a.e. on } \mathbb{T}\setminus J_k.\end{cases}$$

By Lemma 9.3.4 we have $\sup_k\mathcal{D}(h_k)<\infty$. Also h_k^*/f^* is bounded a.e. on J_k (because $d(J_k,E)>0$) and a.e. on $\mathbb{T}\setminus J_k$ (because $|h_k^*/f^*|=|p_kg^*|$ there). Therefore $h_k\in[f]_{\mathcal{D}}$ for all k. Now h_k converges pointwise to pg_n, where p is a monic polynomial of degree $2n$ with zeros in $\mathbb{T}$. By Lemma 9.1.4, $pg_n\in[f]_{\mathcal{D}}$. Using Theorems 9.1.9 and 9.2.1, we have $[pg_n]_{\mathcal{D}}=[p]_{\mathcal{D}}\cap[g_n]_{\mathcal{D}}=[g_n]_{\mathcal{D}}$. Hence, finally, $g_n\in[f]_{\mathcal{D}}$, as claimed. □

The next result is a partial solution to Conjecture 9.3.2. As usual, we write $E_t:=\{\zeta\in\mathbb{T}:d(\zeta,E)\le t\}$, where d is arclength distance on $\mathbb{T}$.

Theorem 9.3.5 *Let $f \in \mathcal{D}$ be an outer function, and define E as in* (9.1). *If*

$$\int_0^{1/e} c(E_t)\frac{\log\log(1/t)}{t\log(1/t)}\,dt < \infty, \tag{9.3}$$

then f is cyclic for $\mathcal{D}$.

Once again, this result falls short of being a full solution to Conjecture 9.3.2 for two reasons. The first, just as before, is that E is defined differently in the theorem and in the conjecture. The second reason is that condition (9.3) is stronger than $c(E) = 0$. Indeed, $c(E) = 0$ is equivalent to $\lim_{t\to 0^+} c(E_t) = 0$ (this follows from Theorem 2.1.6), whereas (9.3) amounts to saying that $c(E_t) \to 0$ at least at a certain rate as $t \to 0^+$.

Theorem 9.3.5 suffers from a further defect. The condition (9.3) is rather difficult to verify in practice. It does hold whenever E is finite (see Exercise 9.3.4), and with some work one can also show that it holds for certain very thin Cantor-type sets, but in general it is rather unwieldy. We shall return to this problem in the next section, replacing the capacity condition by a measure condition that is easier to handle.

For the time being, let us concentrate on the proof of the theorem. The idea is to use (9.3) to construct a 'nice' function g lying in $[f]_{\mathcal{D}}$, and then to deform g into the constant function 1 while staying inside $[f]_{\mathcal{D}}$. The deformation is carried out using the following lemma.

Lemma 9.3.6 *Let $h \in \mathcal{D}$ be a function such that $|\operatorname{Im} h| < \pi/6$ on $\mathbb{D}$. Let $\Omega := \{\lambda \in \mathbb{C} \setminus \{0\} : |\arg(\lambda)| < \pi/6\}$ and, for $\lambda \in \Omega$, define*

$$g_\lambda(z) := \exp(-\lambda e^{h(z)}) \quad (z \in \mathbb{D}).$$

Then $g_\lambda \in \mathcal{D}$ for all $\lambda \in \Omega$, the map $\lambda \mapsto g_\lambda : \Omega \to \mathcal{D}$ is holomorphic, and $\|g_\lambda - 1\|_{\mathcal{D}} \to 0$ as $\lambda \to 0$ from within Ω.

Proof For $\lambda \in \Omega$ and $z \in \mathbb{D}$, we have $|\arg(\lambda e^{h(z)})| < \pi/6 + \pi/6 = \pi/3$. Hence $\operatorname{Re}(\lambda e^h) \geq |\lambda e^h|/2$ on $\mathbb{D}$, and consequently,

$$|g_\lambda| \leq \exp(-|\lambda e^h|/2) \leq 1$$

and

$$|g_\lambda'| \leq |h'||\lambda e^h|\exp(-|\lambda e^h|/2) \leq 2|h'|.$$

It immediately follows that $g_\lambda \in \mathcal{D}$.

If $\lambda_n \to \lambda_0$ in Ω, then $g_{\lambda_n} \to g_{\lambda_0}$ and $g_{\lambda_n}' \to g_{\lambda_0}'$ pointwise on $\mathbb{D}$, and the estimates above allow us to apply the dominated convergence theorem to show that $\|g_{\lambda_n} - g_{\lambda_0}\|_{\mathcal{D}} \to 0$. Thus $\lambda \mapsto g_\lambda$ is continuous as a map : $\Omega \to \mathcal{D}$. The usual argument involving Cauchy's theorem and Morera's theorem then shows that this map is in fact holomorphic.

Finally, let (λ_n) be a sequence in Ω tending to 0. It is clear that $g_{\lambda_n} \to 1$ and $g'_{\lambda_n} \to 0$ pointwise on $\mathbb{D}$. Using the dominated convergence theorem once again, we deduce that $\|g_{\lambda_n} - 1\|_{\mathcal{D}} \to 0$, as required. □

Proof of Theorem 9.3.5 We apply Theorem 3.4.3. Define $\eta(t) := \log\log(1/t)$ for $t \in (0, 1/e)$, and set it equal to zero for $t \geq 1/e$. Then

$$\int_0^{\pi} c(E_t)\,|d\eta^2(t)| = \int_0^{1/e} c(E_t)\frac{2\log\log(1/t)}{t\log(1/t)}\,dt,$$

which is finite by (9.3). Therefore, by Theorem 3.4.3, there exists a function $h \in \mathcal{D}$ such that $|\operatorname{Im} h| < \pi/6$ and

$$\liminf_{z\to\zeta} \operatorname{Re} h(z) \geq \eta(d(\zeta, E)) \qquad (\zeta \in \mathbb{T}).$$

Define Ω and g_λ as Lemma 9.3.6. As in the proof of the lemma, we have $|g_\lambda| \leq \exp(-|\lambda e^h|/2)$ on $\mathbb{D}$. Hence, for each $\lambda \in \Omega$ and $\zeta \in \mathbb{T}$,

$$\limsup_{z\to\zeta} |g_\lambda(z)| \leq \exp(-|\lambda| e^{\eta(d(\zeta,E))}/2) \leq d(\zeta, E)^{|\lambda|/2}.$$

By the maximum principle, it follows that $|g_\lambda(z)| \leq C_\lambda \operatorname{dist}(z, E)^{|\lambda|/2}$, where C_λ is a constant depending on λ (see Exercise 9.3.5). In particular, if $\lambda \in \Omega$ and $|\lambda| \geq 2$, then $|g_\lambda(z)| \leq C_\lambda \operatorname{dist}(z, E)$, so by Theorem 9.3.3 we have $g_\lambda \in [f]_{\mathcal{D}}$. Since $\lambda \mapsto g_\lambda : \Omega \to \mathcal{D}$ is holomorphic, it follows that $g_\lambda \in [f]_{\mathcal{D}}$ for *all* $\lambda \in \Omega$. Finally, since $\|g_\lambda - 1\|_{\mathcal{D}} \to 0$ as $\lambda \to 0$, we obtain $1 \in [f]_{\mathcal{D}}$. Thus f is cyclic. □

Exercises 9.3

1. Let $f \in \mathcal{D}$ be an outer function. Show that

$$\liminf_{z\to\zeta} |(f \wedge 1)(z)| = \min\Bigl\{\liminf_{z\to\zeta} |f(z)|, 1\Bigr\} \qquad (\zeta \in \mathbb{T}).$$

2. The aim of this question is to construct an outer function $f \in \mathcal{D}$ for which the set E defined in (9.1) is the whole unit circle.
   (i) Construct $h \in \mathcal{D}\setminus\{0\}$ whose zeros (w_n) in $\mathbb{D}$ accumulate at every point of $\mathbb{T}$. [Hint: Use Theorem 4.2.1.]
   (ii) Let f be the outer factor of h. Show that $\sum_n |f(w_n)|^2 < \infty$. [Hint: Use Theorem 4.1.3 and the fact that $|h|^2$ is a subharmonic function.]
   (iii) Deduce that $\liminf_{z\to\zeta} |f(z)| = 0$ for all $\zeta \in \mathbb{T}$.
3. Let E be a closed subset of $\mathbb{T}$. Suppose that there exists $g \in \operatorname{Hol}(\mathbb{D}), g \not\equiv 0$ such that $|g(z)| \leq C \operatorname{dist}(z, E)$. Show that E is a Carleson set.

4. Let E be a finite subset of $\mathbb{T}$. Show $c(E_t) \asymp 1/\log(1/t)$ as $t \to 0^+$, and deduce that that (9.3) holds for this E.
5. Let $g \in \mathrm{Hol}(\mathbb{D})$, and suppose that

$$\limsup_{z\to\zeta} |g(z)| \le C d(\zeta, E)^\alpha \qquad (\zeta \in \mathbb{T}),$$

where E is a closed subset E of $\mathbb{T}$, and C, α are positive constants. By applying the maximum principle to $g(z)/(z-a)^\alpha$, where $a \in E$, show that

$$|g(z)| \le C(\pi/2)^\alpha \operatorname{dist}(z, E)^\alpha \qquad (z \in \mathbb{D}).$$

9.4 Measure conditions and distance functions

The criterion (9.3) is expressed in terms of $c(E_t)$, a quantity that is difficult to estimate for all but the simplest sets E. In this section we establish a sufficient condition for cyclicity expressed in terms of $|E_t|$, the Lebesgue measure of E_t, which is somewhat easier to handle.

Theorem 9.4.1 *Let $f \in \mathcal{D}$ be an outer function, and set $E := \{\zeta \in \mathbb{T} : \liminf_{z\to\zeta} |f(z)| = 0\}$. Suppose that, for some $\sigma > 0$,*

$$|E_t| = O(t^\sigma) \quad (t \to 0), \tag{9.4}$$

and that

$$\int_0^\pi \frac{dt}{|E_t|} = \infty. \tag{9.5}$$

Then f is cyclic for $\mathcal{D}$.

The condition (9.4) implies that E is a Carleson set (see Exercise 9.4.1). Condition (9.5) implies that E is of logarithmic capacity zero and, for certain Cantor-type sets, it is actually equivalent to E being of capacity zero (see Exercises 9.4.2 and 9.4.3). Neither of the conditions implies the other.

The basic idea of the proof of Theorem 9.4.1 is the same as that for Theorem 9.3.5, namely to use Theorem 9.3.3 to construct a 'nice' function belonging to $[f]_\mathcal{D}$, and then to deform this function into 1 while staying inside $[f]_\mathcal{D}$. This time, however, the deformation process is somewhat more complicated. It takes place within a certain family of functions called distance functions, which we now proceed to define. In what follows, d denotes the arclength distance on $\mathbb{T}$.

Definition 9.4.2 Let E be a closed subset of $\mathbb{T}$ of Lebesgue measure zero, and let $w : (0, \pi] \to \mathbb{R}^+$ be a continuous function such that

$$\int_{\mathbb{T}} \left| \log w(d(\zeta, E)) \right| |d\zeta| < \infty. \tag{9.6}$$

The *distance function* corresponding to w, E is the outer function $f_{w,E}$ satisfying $|f^*_{w,E}(\zeta)| = w(d(\zeta, E))$ a.e. on $\mathbb{T}$, namely:

$$f_{w,E}(z) := \exp\Big(\frac{1}{2\pi} \int_{\mathbb{T}} \frac{\zeta + z}{\zeta - z} \log w(d(\zeta, E)) \, |d\zeta|\Big) \qquad (z \in \mathbb{D}). \tag{9.7}$$

One of the virtues of distance functions is that, under a certain concavity condition, their Dirichlet integrals are particularly easy to estimate.

Theorem 9.4.3 *Let w, E be as in Definition 9.4.2, and let $f_{w,E}$ be the corresponding distance function. Suppose further that w is increasing, and that there exists $\gamma > 2$ such that $t \mapsto w(t^\gamma)$ is a concave function. Then*

$$\mathcal{D}(f_{w,E}) \leq C_\gamma \int_0^\pi w'(t)^2 |E_t| \, dt, \tag{9.8}$$

where C_γ is a constant depending only on γ.

For the proof, we need an inequality that can be used to check conditions like (9.6).

Lemma 9.4.4 *Let E be a closed subset of $\mathbb{T}$ of Lebesgue measure zero, and let $\phi : \mathbb{T} \to [0, \infty)$ be a positive, measurable function. Then*

$$\int_{\mathbb{T}} \phi(d(\zeta, E)) \, |d\zeta| \leq \int_0^\pi \phi(t) \frac{|E_t|}{t} \, dt.$$

Proof Let (I_j) be the connected components of $\mathbb{T} \setminus E$. Then

$$\int_{\mathbb{T}} \phi(d(\zeta, E)) \, |d\zeta| = \sum_j 2 \int_0^{|I_j|/2} \phi(t) \, dt = \int_0^\pi \phi(t) \psi(t) \, dt, \tag{9.9}$$

where $\psi := 2 \sum_j 1_{[0, |I_j|/2]}$, a positive, decreasing function. In particular, taking $\phi = 1_{[0,\delta]}$, we get $|E_\delta| = \int_0^\delta \psi(t) \, dt \geq \delta \psi(\delta)$, so $\psi(t) \leq |E_t|/t$ for all t. Substituting this inequality back into (9.9) gives the result. □

Proof of Theorem 9.4.3 We use Carleson's formula for the Dirichlet integral of an outer function, Theorem 7.4.1. Let ζ_1, ζ_2 denote points of $\mathbb{T}$, and write

$\delta_j := d(\zeta_j, E)$. In this notation, Carleson's formula becomes

$$\begin{aligned}
\mathcal{D}(f_{w,E}) &= \frac{1}{4\pi^2}\iint_{\mathbb{T}^2} \frac{(w^2(\delta_1) - w^2(\delta_2))(\log w(\delta_1) - \log w(\delta_2))}{|\zeta_1 - \zeta_2|^2} |d\zeta_1|\,|d\zeta_2| \\
&\le \frac{1}{16}\iint_{\mathbb{T}^2} \frac{(w^2(\delta_1) - w^2(\delta_2))(\log w(\delta_1) - \log w(\delta_2))}{d(\zeta_1, \zeta_2)^2} |d\zeta_1|\,|d\zeta_2| \\
&\le \frac{1}{8}\iint_{\delta_1 \ge \delta_2} \frac{(w^2(\delta_1) - w^2(\delta_2))(\log w(\delta_1) - \log w(\delta_2))}{d(\zeta_1, \zeta_2)^2} |d\zeta_1|\,|d\zeta_2|.
\end{aligned}$$

For convenience, let us extend w to the whole of $\mathbb{R}^+$ by defining $w(t) := w(\pi)$ for $t > \pi$. We then have $w(\delta_1) \le w(\delta_2 + d(\zeta_1, \zeta_2))$, and consequently

$$\begin{aligned}
\mathcal{D}(f_{w,E}) &\le \frac{1}{4}\int_{\mathbb{T}}\int_0^{\pi} \frac{(w^2(\delta_2 + s) - w^2(\delta_2))(\log w(\delta_2 + s) - \log w(\delta_2))}{s^2}\, ds\, |d\zeta_2| \\
&\le \frac{1}{4}\int_0^{\pi}\int_0^{\pi} \frac{(w^2(t + s) - w^2(t))(\log w(t + s) - \log w(t))}{s^2}\, ds\, \frac{|E_t|}{t}\, dt,
\end{aligned}$$

where, for the last step, we used Lemma 9.4.4.

To estimate this, we now exploit the concavity assumption on w. This assumption amounts to saying that $t \mapsto w'(t)t^{1-1/\gamma}$ is decreasing. Thus

$$\begin{aligned}
w^2(t + s) - w^2(t) &= \int_t^{t+s} 2w(u)w'(u)\, du \\
&\le \int_t^{t+s} 2w(t + s)w'(t)(t/u)^{1-1/\gamma}\, du \\
&= 2\gamma w(t + s)w'(t)t((1 + s/t)^{1/\gamma} - 1).
\end{aligned}$$

The concavity assumption also implies that $w(t)/t^{1/\gamma}$ is decreasing. Therefore

$$\begin{aligned}
\log w(t + s) - \log w(t) &= \int_t^{t+s} \frac{w'(u)}{w(u)}\, du \\
&= \int_t^{t+s} u^{1-1/\gamma}w'(u)\frac{u^{1/\gamma}}{w(u)}\, \frac{du}{u} \\
&\le \int_t^{t+s} t^{1-1/\gamma}w'(t)\frac{(t + s)^{1/\gamma}}{w(t + s)}\, \frac{du}{u} \\
&= tw'(t)\frac{(1 + s/t)^{1/\gamma}}{w(t + s)} \log(1 + s/t).
\end{aligned}$$

Combining these estimates, we obtain

$$\int_0^\pi \frac{(w^2(t+s) - w^2(t))(\log w(t+s) - \log w(t))}{s^2}\,ds$$
$$\leq \int_0^\pi 2\gamma w'(t)^2 t^2((1+s/t)^{1/\gamma} - 1)(1+s/t)^{1/\gamma}\log(1+s/t)\,\frac{ds}{s^2}$$
$$= w'(t)^2 t \int_0^\infty 2\gamma((1+x)^{1/\gamma} - 1)(1+x)^{1/\gamma}\log(1+x)\,\frac{dx}{x^2}.$$

Since $\gamma > 2$, the integral converges, say to C_γ. Plugging this into the estimate for $\mathcal{D}(f_{w,E})$ yields

$$\mathcal{D}(f_{w,E}) \leq \frac{C_\gamma}{4}\int_0^\pi w'(t)^2|E_t|\,dt,$$

giving the desired result. □

The final ingredient in the proof of Theorem 9.4.1 is a regularization lemma.

Lemma 9.4.5 *Let $a > 0$, let $\beta \in (0,1]$ and let $\phi : (0,a] \to (0,\infty)$ be a function such that*

- *$\phi(t)/t$ is decreasing,*
- *$0 < \phi(t) \leq t^\beta$ for all $t \in (0,a]$,*
- *$\int_0^a dt/\phi(t) = \infty$.*

Then, given $\alpha \in (0,\beta)$, there exists a function $\psi : (0,a] \to (0,\infty)$ such that

- *$\psi(t)/t^\alpha$ is increasing,*
- *$\phi(t) \leq \psi(t) \leq t^\beta$ for all $t \in (0,a]$,*
- *$\int_0^a dt/\psi(t) = \infty$.*

The proof of this result is a real-variable argument involving the so-called rising-sun lemma. It is fairly long so, in order not to interrupt the flow, we postpone it to Appendix D.

Proof of Theorem 9.4.1 Let f be the function in the statement of the theorem. Our aim is to prove that $1 \in [f]_{\mathcal{D}}$.

Let σ be as in the statement of the theorem. We can suppose that $\sigma < 1/2$. Fix α with $1/2 < \alpha < (1+\sigma)/2$, and set $w(t) := t^{1-\alpha}$. By Lemma 9.4.4,

$$\int_{\mathbb{T}} |\log w(d(\zeta, E))|\,|d\zeta| \leq C\int_0^\pi |\log t| t^{\sigma-1}\,dt < \infty,$$

so we may construct the distance function $g := f_w$ (for brevity, we have dropped E from the notation). We claim that $g \in [f]_{\mathcal{D}}$. Indeed, by Theorem 9.4.3, applied with $\gamma = 1/(1-\alpha)$, we have

$$\mathcal{D}(g) \le C \int_0^\pi t^{-2\alpha} t^\sigma \, dt < \infty,$$

so $g \in \mathcal{D}$. As g is bounded, it follows that $g^n \in \mathcal{D}$ for all $n \ge 1$. Also, if n is larger than $1/(1-\alpha)$, then g^n satisfies $|g^n(z)| \le C \operatorname{dist}(z, E)$. Consequently, by Theorem 9.3.3, we have $g^n \in [f]_{\mathcal{D}}$. Also, from Theorem 9.1.7, we have $[g^n]_{\mathcal{D}} = [g]_{\mathcal{D}}$, and thus $g \in [f]_{\mathcal{D}}$, as claimed.

The rest of the proof consists of showing that $1 \in [g]_{\mathcal{D}}$. We shall achieve this by constructing a family of functions $w_\delta : (0, \pi] \to \mathbb{R}^+$ for $0 < \delta < 1$, such that the corresponding distance functions f_{w_δ} belong to $[g]_{\mathcal{D}}$, converge pointwise to 1 as $\delta \to 0$, and satisfy $\liminf_{\delta \to 0} \mathcal{D}(f_{w_\delta}) < \infty$. If such a family exists, then by Lemma 9.1.4 we have $1 \in [g]_{\mathcal{D}}$, as desired.

Here is the construction. Fix β with $\alpha < \beta < (1+\sigma)/2$, and define a function $\phi : (0, \pi] \to \mathbb{R}^+$ by $\phi(t) := \min\{|E_t|, t^\beta\}$. This function satisfies the hypotheses of the regularization lemma, Lemma 9.4.5, so there exists a function $\psi : (0, \pi] \to \mathbb{R}^+$ satisfying the conclusions of that lemma, namely: $\psi(t)/t^\alpha$ is increasing, $\phi(t) \le \psi(t) \le t^\beta$ for all $t \in (0, \pi]$, and $\int_0^\pi dt/\psi(t) = \infty$. Note that, for $0 < t < 1$, we have $\psi(t) \ge \phi(t) \ge t$. For $0 < \delta < 1$, define $w_\delta : (0, \pi] \to \mathbb{R}^+$ by

$$w_\delta(t) := \begin{cases} \delta^\alpha t^{1-\alpha}/\psi(\delta), & 0 < t \le \delta, \\ A_\delta - \log \int_t^\pi ds/\psi(s), & \delta < t \le \eta_\delta, \\ 1, & \eta_\delta < t \le \pi. \end{cases}$$

Here A_δ, η_δ are constants, chosen to make w_δ a continuous function satisfying $0 \le w_\delta \le 1$ (see Figure 9.1).

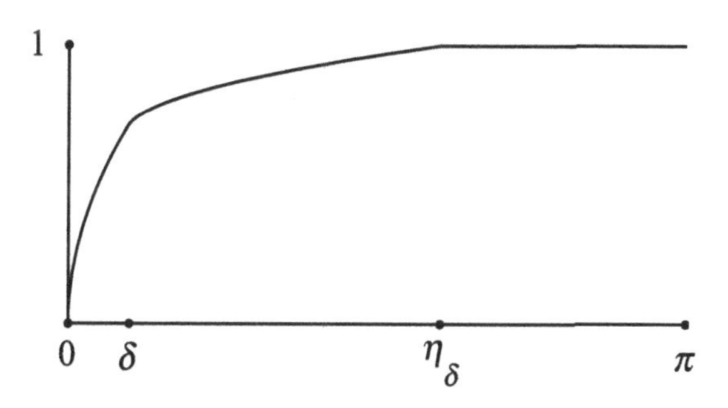

Figure 9.1 The function w_δ

Just as before, w_δ satisfies (9.6), and so we may define the corresponding distance function f_{w_δ}. We now check that f_{w_δ} has the required properties.

We claim that there exists $\gamma > 2$ such that, for all sufficiently small $\delta > 0$, the function $t \mapsto w_\delta(t^\gamma)$ is concave on $(0, \pi]$. Assume this for the moment. Then Theorem 9.4.3 applies and, for all small δ, we have $f_{w_\delta} \in \mathcal{D}$. Moreover, since $w_\delta(t)/t^{1-\alpha}$ is bounded, so too is f_{w_δ}/g. Therefore, by Theorem 9.1.2, $f_{w_\delta} \in [g]_\mathcal{D}$.

The fact that $f_{w_\delta} \to 1$ pointwise as $\delta \to 0$ is an easy consequence of the assertion that $\lim_{\delta\to 0} \eta_\delta = 0$, which we now prove. Given $\epsilon > 0$, if $\eta_\delta > \epsilon$, then $w_\delta(\epsilon) < 1$, in other words

$$\log \int_\delta^\pi \frac{ds}{\psi(s)} - \frac{\delta}{\psi(\delta)} - \log \int_\epsilon^\pi \frac{ds}{\psi(s)} < 1.$$

As $\delta \to 0$, the left-hand side tends to infinity. Thus $\eta_\delta \le \epsilon$ for all sufficiently small δ.

The last property to be checked is that $\liminf_{\delta\to 0} \mathcal{D}(f_{w_\delta}) = 0$. This requires a bit more work. By Theorem 9.4.3 again, for all δ small enough,

$$\mathcal{D}(f_{w_\delta}) \le C \int_0^\pi w_\delta'(t)^2 |E_t|\, dt,$$

where C is a constant independent of δ. We examine this integral separately on $(0, \delta)$ and (δ, η_δ).

Let us begin with (δ, η_δ). Here we have

$$\int_\delta^{\eta_\delta} w_\delta'(t)^2 |E_t|\, dt = \int_\delta^{\eta_\delta} \Big(\int_t^\pi \frac{ds}{\psi(s)} \Big)^{-2} \frac{|E_t|}{\psi(t)^2}\, dt.$$

Note that if $|E_t| \le t^\beta$ then $\psi(t) \ge |E_t|$, whereas if $|E_t| > t^\beta$ then $\psi(t) = t^\beta$. The last integral is therefore majorized by

$$\begin{aligned} &\int_\delta^{\eta_\delta} \Big(\int_t^\pi \frac{ds}{\psi(s)} \Big)^{-2} \frac{1}{\psi(t)}\, dt + \int_\delta^{\eta_\delta} \Big(\int_t^\pi \frac{ds}{\psi(s)} \Big)^{-2} \frac{Ct^\sigma}{t^{2\beta}}\, dt \\ &\qquad \le \Big(\int_{\eta_\delta}^\pi \frac{ds}{\psi(s)} \Big)^{-1} + C\Big(\int_{\eta_\delta}^\pi \frac{ds}{\psi(s)} \Big)^{-2} \eta_\delta^{\sigma+1-2\beta}, \end{aligned}$$

and this tends to zero as $\delta \to 0$.

Now we consider what happens on $(0, \delta)$. Here we have

$$\int_0^\delta w_\delta'(t)^2 |E_t|\, dt = \frac{\delta^{2\alpha}}{\psi(\delta)^2} \int_0^\delta t^{-2\alpha} |E_t|\, dt.$$

If $|E_t| \le t^\beta$ for all $t \in (0, \delta)$, then $|E_t|/t^\alpha \le \psi(t)/t^\alpha \le \psi(\delta)/\delta^\alpha$, and so,

$$\frac{\delta^{2\alpha}}{\psi(\delta)^2} \int_0^\delta t^{-2\alpha} |E_t|\, dt \le \frac{\delta^\alpha}{\psi(\delta)} \int_0^\delta t^{-\alpha}\, dt = \frac{1}{1-\alpha} \frac{\delta}{\psi(\delta)} \le \frac{1}{1-\alpha}.$$

On the other hand, if $|E_t| > t^\beta$ for a sequence $t = \delta_n$ tending to zero, then $\psi(\delta_n) = \delta_n^\beta$ for all n, and consequently

$$\frac{\delta_n^{2\alpha}}{\psi(\delta_n)^2}\int_0^{\delta_n} t^{-2\alpha}|E_t|\,dt \le \frac{\delta_n^{2\alpha}}{\delta_n^{2\beta}}\int_0^{\delta_n} t^{-2\alpha}Ct^\sigma\,dt = \frac{C}{1+\sigma-2\alpha}\delta_n^{1+\sigma-2\beta},$$

which tends to zero as $n \to \infty$.

Putting everything together, we get $\liminf_{\delta\to 0}\mathcal{D}(f_{w_\delta}) = 0$, as we wanted.

All that remains is to establish the claim about concavity. Fix $\gamma > 2$ with $1 - 1/\gamma < \alpha$. Our aim is to prove that $t^{1-1/\gamma}w_\delta'(t)$ is decreasing. On $(0,\delta)$ we have

$$t^{1-1/\gamma}w_\delta'(t) = Ct^{-\nu},$$

where $\nu := \alpha + 1/\gamma - 1 > 0$. This is certainly decreasing. On (δ, η_δ) we have

$$t^{1-1/\gamma}w_\delta'(t) = \frac{t^{1-1/\gamma}}{\psi(t)}\Big(\int_t^\pi \frac{ds}{\psi(s)}\Big)^{-1} = \frac{t^\alpha}{\psi(t)}\Big(t^\nu\int_t^\pi \frac{ds}{\psi(s)}\Big)^{-1}.$$

Now $\psi(t)/t^\alpha$ is increasing. Also, the derivative of $t \mapsto t^\nu\int_t^\pi ds/\psi(s)$ has the same sign as

$$\nu\int_t^\pi \frac{ds}{\psi(s)} - \frac{t}{\psi(t)},$$

which is positive if t is small enough. Thus $t^{1-1/\gamma}w_\delta'(t)$ is decreasing on (δ,η_δ) provided that δ is small enough. Lastly, at $t = \delta$, we need that the left derivative of w_δ exceeds the right derivative, which boils down to the inequality

$$\int_\delta^\pi \frac{ds}{\psi(s)} \ge \frac{1}{1-\alpha},$$

and this certainly holds for small δ, since the left-hand side tends to infinity as $\delta \to 0$. In summary, we have shown that $t^{1-1/\gamma}w_\delta'(t)$ is decreasing on $(0,\pi]$ if δ is small enough. The claim about concavity is proved, and with it, the theorem. □

Exercises 9.4

1. Let E be a closed subset of $\mathbb{T}$ of measure zero.
   (i) Let $\phi : (0,\pi] \to \mathbb{R}^+$ be a positive, decreasing, differentiable function. Show that
   $$\int_{\mathbb{T}} \phi(d(\zeta,E))\,|d\zeta| = \int_0^\pi |\phi'(t)||E_t|\,dt + 2\pi\phi(\pi).$$
   (ii) Deduce that E is a Carleson set if and only if $\int_0^\pi (|E_t|/t)\,dt < \infty$.
2. Let E be a closed subset of $\mathbb{T}$.

(i) Let $N_E(t)$ be the *t-covering number* of E, namely the number of sets of diameter t needed to cover E, where diameter is measured using the arclength metric. Show that

$$(t/2)N_E(t) \leq |E_t| \leq 2tN_E(t) \qquad (0 < t \leq \pi).$$

[Hint for the left-hand inequality: Consider a maximal disjoint collection of closed arcs of diameter $t/2$ all of which meet E.]

(ii) Deduce that the logarithmic capacity $c(E)$ satisfies

$$\frac{1}{c(E)} \geq \int_0^\pi \frac{1}{2|E_t|}\,dt.$$

[Hint: Use Theorem 2.3.2, taking into account the fact that logarithmic capacity is defined using the chordal metric on $\mathbb{T}$ rather than the arclength metric.]

3. Let $(l_n)_{n\geq 0}$ be a sequence in $(0, 2\pi)$ such that $\lambda := \sup_{n\geq 0}(l_{n+1}/l_n) < 1/2$, and let E be the associated circular Cantor set. (Thus, we begin with a closed arc of length l_0, remove an open arc from the middle to leave two closed arcs of length l_1, remove open arcs from their middles to leave four arcs of length l_2, etc.; then E is the intersection of the resulting nested sequence of sets.)

(i) Show that $|E_t| = O(t^\sigma)$ as $t \to 0$, where $\sigma := 1 - \log 2/\log(1/\lambda)$.

(ii) Show that the following statements are equivalent:

- $c(E) = 0$,
- $\sum_{n\geq 0} 2^{-n}\log(1/l_n) = \infty$,
- $\int_0^\pi dt/|E_t| = \infty$.

[Hint: Use the preceding exercise and Theorem 2.3.5.]

(iii) Deduce that, if $f \in \mathcal{D} \cap A(\mathbb{D})$ vanishes precisely on E, then f is cyclic for $\mathcal{D}$ if and only if f is outer and $c(E) = 0$.

9.5 Cyclicity via duality

By the Hahn–Banach theorem, if the only element of the dual space of $\mathcal{D}$ that vanishes on $[f]_\mathcal{D}$ is zero, then $[f]_\mathcal{D} = \mathcal{D}$ and f is cyclic for $\mathcal{D}$. This suggests analyzing further those elements of the dual space of $\mathcal{D}$ that vanish on $[f]_\mathcal{D}$, which is what we are going to do in this section.

There are various ways of representing the dual of $\mathcal{D}$. One, of course, is as $\mathcal{D}$ itself, via the usual Hilbert-space duality, defined using the (sesquilinear) inner product $\langle\cdot,\cdot\rangle_\mathcal{D}$. However, we shall instead identify the dual of $\mathcal{D}$ with a certain

Bergman-type space $\mathcal{B}_e$, using a complex bilinear pairing $\langle \cdot, \cdot \rangle$. We begin by describing this pairing in detail. In what follows, $\mathbb{C}_\infty$ denotes the Riemann sphere.

Definition 9.5.1 Let $\mathbb{D}_e := \{z \in \mathbb{C}_\infty : |z| > 1\}$. We write $\mathcal{B}_e$ for the space of all $\phi \in \text{Hol}(\mathbb{D}_e)$ such that

$$\|\phi\|_{\mathcal{B}_e}^2 := \frac{1}{\pi} \int_{\mathbb{D}_e} |\phi(z)|^2 \frac{dA(z)}{|z|^2} < \infty.$$

We can express $\|\phi\|_{\mathcal{B}_e}$ in terms of the Laurent coefficients of ϕ as follows.

Proposition 9.5.2 *Let $\phi \in \text{Hol}(\mathbb{D}_e)$. If $\phi(\infty) \neq 0$, then $\|\phi\|_{\mathcal{B}_e} = \infty$. If $\phi(\infty) = 0$, say $\phi(z) = \sum_{k \geq 0} b_k / z^{k+1}$, then*

$$\|\phi\|_{\mathcal{B}_e}^2 = \sum_{k \geq 0} \frac{|b_k|^2}{k+1}. \tag{9.10}$$

Proof The first statement follows from the fact that $dA(z)/|z|^2$ is an infinite measure on $\mathcal{D}_e$. For the second, writing the area integral in polar coordinates, we have

$$\begin{aligned} \|\phi\|_{\mathcal{B}_e}^2 &= \frac{1}{\pi} \int_{\mathbb{D}_e} \Big| \sum_{k \geq 0} \frac{b_k}{z^{k+1}} \Big|^2 \frac{dA(z)}{|z|^2} \\ &= \frac{1}{\pi} \int_1^\infty \int_0^{2\pi} \Big| \sum_{k \geq 0} b_k r^{-k-1} e^{-i(k+1)\theta} \Big|^2 \, d\theta \, \frac{dr}{r}. \end{aligned}$$

By Parseval's formula, for each $r \in (0, 1)$,

$$\frac{1}{2\pi} \int_0^{2\pi} \Big| \sum_{k \geq 0} b_k r^{-k-1} e^{-i(k+1)\theta} \Big|^2 \, d\theta = \sum_{k \geq 0} |b_k|^2 r^{-2k-2}.$$

Hence

$$\|\phi\|_{\mathcal{B}_e}^2 = 2 \int_1^\infty \sum_{k \geq 0} |b_k|^2 r^{-2k-3} \, dr = \sum_{k \geq 0} \frac{|b_k|^2}{k+1}. \qquad \square$$

This result immediately shows that $(\mathcal{B}_e, \| \cdot \|_{\mathcal{B}_e})$ is a Hilbert space. Also, the appearance of the $k+1$ in the denominator in (9.10) strongly suggests defining the following pairing.

Definition 9.5.3 If $f(z) = \sum_{k \geq 0} a_k z^k \in \mathcal{D}$ and $\phi(z) = \sum_{k \geq 0} b_k / z^{k+1} \in \mathcal{B}_e$, then we write

$$\langle f, \phi \rangle := \sum_{k \geq 0} a_k b_k.$$

Proposition 9.5.4 *For all $f \in \mathcal{D}$ and $\phi \in \mathcal{B}_e$, the pairing $\langle f, \phi \rangle$ is well defined, and*

$$|\langle f, \phi \rangle| \leq \|f\|_{\mathcal{D}} \|\phi\|_{\mathcal{B}_e}.$$

Proof By the Cauchy–Schwarz inequality,

$$\sum_{k \geq 0} |a_k b_k| \leq \Big(\sum_{k \geq 0} (k+1)|a_k|^2 \Big)^{1/2} \Big(\sum_{k \geq 0} |b_k|^2/(k+1) \Big)^{1/2}. \qquad \square$$

In fact it is easy to see that the pairing $\langle \cdot, \cdot \rangle$ establishes an isometric isomorphism between $\mathcal{B}_e$ and the dual of $\mathcal{D}$. This prompts the following definition.

Definition 9.5.5 Given $S \subset \mathcal{D}$, we write

$$S^{\perp} := \{\phi \in \mathcal{B}_e : \langle f, \phi \rangle = 0 \text{ for all } f \in S\}.$$

By the Hahn–Banach theorem, f is cyclic for $\mathcal{D}$ if and only if $[f]_{\mathcal{D}}^{\perp} = \{0\}$. We are therefore interested in calculating $[f]_{\mathcal{D}}^{\perp}$. The following theorem is the central result of this section. Recall that $A(\mathbb{D})$ denotes the disk algebra and H^p denotes the Hardy spaces on $\mathbb{D}$.

Theorem 9.5.6 *Let $f \in \mathcal{D} \cap A(\mathbb{D})$, and define $E := \{z \in \overline{\mathbb{D}} : f(z) = 0\}$. Let $\phi \in \mathcal{B}_e$, and suppose that $\phi \in [f]_{\mathcal{D}}^{\perp}$. Then*

(i) *ϕ extends to be holomorphic in $\mathbb{C}_\infty \setminus E$,*
(ii) *$f\phi$ extends to be holomorphic on $\mathbb{D}$, and $f\phi \in \cap_{0<p<1} H^p$.*

Proof (i) The first step is to establish a formula for $\phi(\lambda)$ in terms of the action of ϕ as a linear functional on $\mathcal{D}$. The formula is

$$\phi(\lambda) = \langle (\lambda - u)^{-1}, \phi \rangle \qquad (|\lambda| > 1), \tag{9.11}$$

where u denotes the function $u(z) := z$. To prove it, write $\phi(z) = \sum_{k \geq 0} b_k / z^{k+1}$. Then, for $|\lambda| > 1$, we have

$$\langle (\lambda - u)^{-1}, \phi \rangle = \Big\langle \sum_{k \geq 0} u^k / \lambda^{k+1}, \phi \Big\rangle = \sum_{k \geq 0} \langle u^k, \phi \rangle / \lambda^{k+1} = \sum_{k \geq 0} b_k / \lambda^{k+1} = \phi(\lambda),$$

the passage to the limit in the second equality justified by Proposition 9.5.4.

The next step is to extend ϕ from $\mathcal{D}_e$ to a larger domain. For this, we use some abstract functional analysis. Observe that $\mathcal{D} \cap A(\mathbb{D})$ is a Banach algebra with respect to the norm $\|h\|_{\mathcal{D} \cap A(\mathbb{D})} := \mathcal{D}(h)^{1/2} + \|h\|_{A(\mathbb{D})}$. Let I be the closed ideal in $\mathcal{D} \cap A(\mathbb{D})$ generated by f. Every element of I is a limit in $\mathcal{D} \cap A(\mathbb{D})$ (and hence in $\mathcal{D}$) of a sequence $(p_n f)$, where p_n are polynomials. It follows that $I \subset [f]_{\mathcal{D}}$. Since $\phi \in [f]_{\mathcal{D}}^{\perp}$, it vanishes on I, and therefore it induces a continuous linear functional $\widetilde{\phi}$ on the quotient algebra $(\mathcal{D} \cap A(\mathbb{D}))/I$ via the formula

$$\langle \widetilde{h}, \widetilde{\phi} \rangle := \langle h, \phi \rangle \qquad (h \in \mathcal{D} \cap A(\mathbb{D})).$$

(Here $\widetilde{h}$ is shorthand for the coset $h + I$. Also, we are abusing notation slightly in continuing to use $\langle \cdot, \cdot \rangle$ for this quotient pairing.) In conjunction with (9.11), this gives

$$\phi(\lambda) = \langle (\lambda - \widetilde{u})^{-1}, \widetilde{\phi} \rangle \qquad (|\lambda| > 1). \tag{9.12}$$

Notice that the right-hand side of this last formula defines a holomorphic continuation of ϕ to the whole of $\mathbb{C}_\infty \setminus \sigma(\widetilde{u})$, where $\sigma(\widetilde{u})$ denotes the spectrum of $\widetilde{u}$ in the (Banach) quotient algebra $(\mathcal{D} \cap A(\mathbb{D}))/I$.

The last step is to show that $\sigma(\widetilde{u}) \subset E$. Let $\lambda \in \sigma(\widetilde{u})$. By the Gelfand theory, there exists a character (multiplicative linear functional) χ on $(\mathcal{D} \cap A(\mathbb{D}))/I$ such that $\chi(\widetilde{u}) = \lambda$. Then, for any polynomial p, we have

$$\chi(\widetilde{p}) = \chi(p(\widetilde{u})) = p(\chi(\widetilde{u})) = p(\lambda).$$

Since polynomials are dense in $\mathcal{D} \cap A(\mathbb{D})$, it follows that $\chi(\widetilde{f}) = f(\lambda)$. But $\widetilde{f} = 0$, since $f \in I$. Consequently $f(\lambda) = 0$, and so $\lambda \in E$, as claimed. This completes the proof of part (i).

(ii) The basis for this part of the proof is another formula for ϕ, or more precisely for $f\phi$, but this time inside $\mathbb{D}$. The formula is

$$f(\lambda)\phi(\lambda) = \langle f, (\lambda - u)^{-1}\phi \rangle \qquad (\lambda \in \mathbb{D} \setminus E). \tag{9.13}$$

Notice that $\mathcal{B}_e$ is stable under multiplication by bounded holomorphic functions on $\mathbb{D}_e$, so $(\lambda - u)^{-1}\phi \in \mathcal{B}_e$ if $|\lambda| < 1$, and the formula at least makes sense. Also, once (9.13) is established, the right-hand side clearly provides a holomorphic continuation of $f\phi$ to the whole of $\mathbb{D}$.

To prove (9.13), we start from (9.12). Fix $\lambda \in \mathbb{D} \setminus E$. We shall construct an explicit inverse for $(\widetilde{u} - \lambda)$ in $(\mathcal{D} \cap A(\mathbb{D}))/I$. Define

$$g(z) := \frac{f(z) - f(\lambda)}{z - \lambda} \qquad (z \in \mathbb{D} \setminus \{\lambda\}).$$

Clearly λ is a removable singularity and, after its removal, $g \in \mathcal{D} \cap A(\mathbb{D})$. Also $(u - \lambda)g = f - f(\lambda)$, so in the quotient algebra $(\widetilde{u} - \lambda)\widetilde{g} = -f(\lambda)$, and hence $f(\lambda)(\widetilde{u} - \lambda)^{-1} = -\widetilde{g}$. Plugging this into (9.12), we obtain

$$f(\lambda)\phi(\lambda) = -\langle \widetilde{g}, \widetilde{\phi} \rangle = -\langle g, \phi \rangle.$$

We now derive an expansion for g in powers of λ. Define $T : \mathcal{D} \to \mathcal{D}$ by

$(Th)(z) := (h(z) - h(0))/z$. Clearly T acts as a contraction on $\mathcal{D}$. Now

$$\begin{aligned}(u-\lambda)\sum_{k\geq 0}\lambda^k T^{k+1} f &= f + \sum_{k\geq 0}\lambda^k (uT^{k+1}f - T^k f)\\ &= f - \sum_{k\geq 0}\lambda^k (T^k f)(0)\\ &= f - f(\lambda),\end{aligned}$$

the last equation arising from the fact that $(T^k f)(0)$ is the k-th Taylor coefficient of f. A comparison with the definition of g reveals that $g = \sum_{k\geq 0}\lambda^k T^{k+1} f$. Substituting this expression for g into the previous formula for $f(\lambda)\phi(\lambda)$, we get

$$f(\lambda)\phi(\lambda) = -\sum_{k\geq 0}\lambda^k\langle T^{k+1}f,\phi\rangle = -\sum_{k\geq 0}\lambda^k\langle f,\phi/u^{k+1}\rangle = \langle f,(\lambda-u)^{-1}\phi\rangle.$$

This establishes (9.13) and, as already remarked, shows that $f\phi\in \mathrm{Hol}(\mathbb{D})$.

It remains to prove that $f\phi\in\cap_{0<p<1}H^p$. For this we use a little more functional analysis. Let $A(\mathbb{D}_e)$ denote the disk algebra on $\mathbb{D}_e$, namely the algebra of functions continuous on $\overline{\mathbb{D}}_e$ and holomorphic on $\mathbb{D}_e$. It is a Banach algebra with respect to the sup-norm on $\overline{\mathbb{D}}_e$ (or equivalently on $\mathbb{T}$). If $h\in A(\mathbb{D}_e)$, then clearly $h\phi\in\mathcal{B}_e$ and $\|h\phi\|_{\mathcal{B}_e}\leq\|h\|_{A(\mathbb{D}_e)}\|\phi\|_{\mathcal{B}_e}$. Consequently

$$|\langle f,h\phi\rangle|\leq\|f\|_{\mathcal{D}}\|\|\phi\|_{\mathcal{B}_e}\|h\|_{A(\mathbb{D}_e)}\qquad(h\in A(\mathbb{D}_e)).$$

This implies that $h\mapsto\langle f,h\phi\rangle$ is a continuous linear functional on $A(\mathbb{D}_e)$. Thus there exists a complex measure μ on $\mathbb{T}$ such that

$$\langle f,h\phi\rangle = \int_{\mathbb{T}} h(\zeta)\,d\mu(\zeta)\qquad(h\in A(\mathbb{D}_e)).$$

In particular, taking $h=(\lambda-u)^{-1}$, and using formula (9.13), we obtain

$$f(\lambda)\phi(\lambda) = \int_{\mathbb{T}}\frac{d\mu(\zeta)}{\lambda-\zeta}\qquad(\lambda\in\mathbb{D}).$$

In other words, $f\phi$ is the Cauchy transform of $-\mu$ and, as such, automatically belongs to $\cap_{0<p<1}H^p$ (see Corollary A.1.12 in Appendix A). This completes the proof of part (ii) of the theorem. □

Of particular interest is the case where f is outer. Recall that $\mathcal{N}^+$ denotes the Smirnov class, namely the set of functions holomorphic in $\mathbb{D}$ expressible as the product of an inner function and an outer function. (For more information about the Smirnov class, see Appendix A.3.)

Corollary 9.5.7 *Let $f\in\mathcal{D}\cap A(\mathbb{D})$ be outer and define $E:=\{z\in\mathbb{T}: f(z)=0\}$. If $\phi\in[f]_{\mathcal{D}}^{\perp}$, then*

(i) *ϕ extends to be holomorphic on $\mathbb{C}_\infty \setminus E$,*
(ii) *$\phi|_{\mathbb{D}_e} \in \mathcal{B}_e$ and $\phi|_{\mathbb{D}} \in \mathcal{N}^+$.*

Proof By Theorem 9.5.6, the function ϕ extends to be holomorphic on $\mathbb{C}_\infty \setminus E$ (note that f, being outer, has no zeros inside $\mathbb{D}$). Clearly $\phi|_{\mathbb{D}_e} \in \mathcal{B}_e$. Also, we have $\phi|_{\mathbb{D}} = (f\phi)/f$, where the numerator belongs to $\cap_{0<p<1} H^p$, and the denominator is outer. Consequently $\phi|_{\mathbb{D}} \in \mathcal{N}^+$. □

Finally, we deduce a cyclicity result of the type that we are seeking.

Corollary 9.5.8 *Let $f \in \mathcal{D} \cap A(\mathbb{D})$ be outer and define $E := \{z \in \mathbb{T} : f(z) = 0\}$. Suppose that E has the property that*

$$\left.\begin{array}{l} \phi \in \mathrm{Hol}(\mathbb{C}_\infty \setminus E) \\ \phi|_{\mathbb{D}_e} \in \mathcal{B}_e \\ \phi|_{\mathbb{D}} \in \mathcal{N}^+ \end{array}\right\} \quad \Rightarrow \quad \phi \equiv 0. \tag{9.14}$$

Then f is cyclic for $\mathcal{D}$.

Proof By Corollary 9.5.7 and the assumption (9.14), we have $[f]^{\perp}_{\mathcal{D}} = \{0\}$. By the Hahn–Banach theorem, f is cyclic for $\mathcal{D}$. □

This result still begs the question as to which sets E satisfy (9.14). This is the subject of the final section of the chapter.

9.6 Bergman–Smirnov exceptional sets

In the light of the previous result, it is natural to make the following definition.

Definition 9.6.1 Let E be a closed subset of $\mathbb{T}$. We write

$$\mathrm{Hol}_{\mathcal{B}_e,\mathcal{N}^+}(\mathbb{C}_\infty \setminus E) := \{\phi \in \mathrm{Hol}(\mathbb{C}_\infty \setminus E) : \phi|_{\mathcal{D}_e} \in \mathcal{B}_e,\ \phi|_{\mathbb{D}} \in \mathcal{N}^+\}.$$

We say that E is a *Bergman–Smirnov exceptional set* if $\mathrm{Hol}_{\mathcal{B}_e,\mathcal{N}^+}(\mathbb{C}_\infty \setminus E) = \{0\}$.

Using this terminology, we can restate Corollary 9.5.8 more succinctly as follows.

Theorem 9.6.2 *Let $f \in \mathcal{D} \cap A(\mathbb{D})$ be outer and define $E := \{z \in \mathbb{T} : f(z) = 0\}$. If E is a Bergman–Smirnov exceptional set, then f is cyclic for $\mathcal{D}$.* □

Thus the problem of determining whether a function is cyclic for $\mathcal{D}$ is transformed into an apparently quite different problem, namely that of characterizing the Bergman–Smirnov exceptional sets. We shall now attempt to do this.

We begin by showing that Bergman–Smirnov exceptional sets cannot be too large. Recall that $c(E)$ denotes the logarithmic capacity of E.

Theorem 9.6.3 *If E is a Bergman–Smirnov exceptional set, then $c(E) = 0$.*

Proof Suppose that $c(E) > 0$. Then there exists a probability measure μ on E with finite energy $I_K(\mu)$. Define

$$\phi(z) := \int_E \frac{\zeta\, d\mu(\zeta)}{\zeta - z} \qquad (z \in \mathbb{C} \setminus E).$$

Clearly ϕ is holomorphic in $\mathbb{C} \setminus E$, with a removable singularity at infinity. Also, as ϕ is the Cauchy transform of a finite measure, $\phi|_{\mathbb{D}} \in \cap_{0<p<1} H^p \subset \mathcal{N}^+$ (see Corollary A.1.12 in Appendix A). Further, on $\mathbb{D}_e$, we have

$$\phi(z) = -\sum_{k\geq 0} \int_E (\zeta^{k+1}/z^{k+1})\, d\mu(\zeta) = -\sum_{k\geq 0} \widehat{\mu}(k+1)/z^{k+1} \qquad (|z| > 1),$$

and so by Proposition 9.5.2 and then Theorem 2.4.4,

$$\|\phi\|^2_{\mathcal{B}_e} = \sum_{k\geq 0} \frac{|\widehat{\mu}(k+1)|^2}{k+1} = I_K(\mu) - \log 2 < \infty.$$

Therefore $\phi|_{\mathbb{D}_e} \in \mathcal{B}_e$. Putting all these observations together, we have $\phi \in \mathrm{Hol}_{\mathcal{B}_e,\mathcal{N}^+}(\mathbb{C}_\infty \setminus E)$. Finally, $\phi \not\equiv 0$, because $\phi(0) = \int_E d\mu = 1$. We conclude that E is not a Bergman–Smirnov exceptional set. □

We now seek examples of Bergman–Smirnov exceptional sets. Clearly, the smaller E is, the more likely it is to be Bergman–Smirnov exceptional. The first candidate is thus the empty set. This is indeed Bergman–Smirnov exceptional because, if $\phi \in \mathrm{Hol}(\mathbb{C}_\infty)$, then ϕ is constant, and if further $\phi|_{\mathcal{D}_e} \in \mathcal{B}_e$, then the constant must be zero.

The next case is when E is a singleton. This too turns out to be Bergman–Smirnov exceptional, but it is already harder to prove. To do so, we shall first establish some general properties of the space $\mathrm{Hol}_{\mathcal{B}_e,\mathcal{N}^+}(\mathbb{C}_\infty \setminus E)$, beginning with the following estimate.

Theorem 9.6.4 *If $\phi \in \mathrm{Hol}_{\mathcal{B}_e,\mathcal{N}^+}(\mathbb{C}_\infty \setminus E)$, then there is a constant C such that*

$$|\phi(z)| \leq \frac{C}{\mathrm{dist}(z, E)^2} \qquad (1 < |z| < 2). \tag{9.15}$$

This estimate is based on a two-sided maximum principle.

Lemma 9.6.5 *Let $h \in \mathrm{Hol}(\mathbb{D})$, and suppose that there exists a constant $A > 0$ such that*

$$|h(x+iy)| \leq \begin{cases} 1/y, & y > 0, \\ \exp(A/|y|), & y < 0. \end{cases} \tag{9.16}$$

Then

$$|h(z)| \leq \frac{8e^A}{|1-z^2|^2} \qquad (z \in \mathbb{D},\ \operatorname{Im} z \geq 0).$$

Proof We may assume that h is holomorphic even in a neighborhood of $\overline{\mathbb{D}}$. The general case follows by considering $rh(rz)$ and letting $r \to 1^-$.

Let Γ_1 be the quarter of the circle $|z-(1+i)| = 1$ joining i to 1, and let Γ_2 be its reflection in the y-axis. Also let Γ_3 be the lower half of the circle $|z| = 1$. Finally, let Ω be the region bounded by the three arcs Γ_1, Γ_2 and Γ_3 (see Figure 9.2).

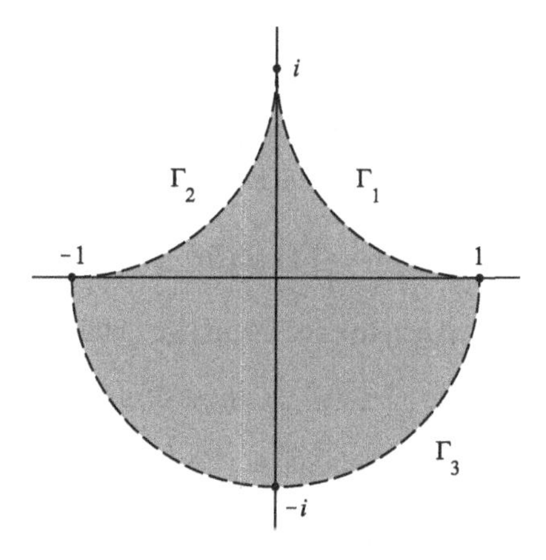

Figure 9.2 The region Ω

We are going to apply the maximum principle to the function g defined on Ω by

$$g(z) := h(z)(1-z^2)^2 \exp\Bigl(-\frac{2Aiz}{1-z^2}\Bigr) \qquad (z \in \Omega).$$

Clearly g is holomorphic in Ω, and extends continuously to every point of $\overline{\Omega}$, except possibly ± 1. In fact, defining $g(\pm 1) := 0$ makes g continuous even at these points. To see this, observe that $\operatorname{Im}(1/(1-z)) < 1/2$ outside the circle containing Γ_1 and $\operatorname{Im}(1/(-1-z)) < 1/2$ outside the circle containing Γ_2, so

$$\Bigl|\exp(-\frac{2Aiz}{1-z^2})\Bigr| = \exp\Bigl(A \operatorname{Im}\Bigl(\frac{1}{1-z} + \frac{1}{-1-z}\Bigr)\Bigr) \leq e^A \qquad (z \in \Omega).$$

Hence $|g(z)| \leq (\sup_{\overline{\mathbb{D}}} |h|)|1-z^2|^2 e^A$, which tends to zero as $z \to \pm 1$ in Ω.

We now estimate $|g|$ on the boundary of Ω. A simple calculation shows that, on Γ_1, we have $|1-z|^2 = 2y$ and $|1+z| \le 2$, so

$$|g(z)| \le \frac{1}{y}|1-z^2|^2 e^A \le 8e^A \qquad (z \in \Gamma_1).$$

By symmetry, the same estimate holds on Γ_2. Also, on the lower half of the unit circle, we have

$$|g(e^{-i\theta})| \le \exp\Bigl(\frac{A}{\sin\theta}\Bigr)(2\sin\theta)^2 \exp\Bigl(-\frac{A}{\sin\theta}\Bigr) \le 4 \qquad (0<\theta<\pi),$$

so $|g| \le 4$ on Γ_3. Applying the maximum principle, we deduce that $|g| \le 8e^A$ on Ω.

In particular, $|g| \le 8e^A$ on the interval $(-1,1)$. This implies that

$$|h(z)(1-z^2)^2| \le 8e^A \qquad (z \in [-1,1]).$$

The same estimate also holds on the upper half of the unit circle. Indeed, re-using the initial hypothesis on h, we have

$$|h(e^{i\theta})(1-e^{2i\theta})^2| \le \Bigl(\frac{1}{\sin\theta}\Bigr)^2(2\sin\theta)^2 \le 4 \qquad (0<\theta<\pi).$$

Applying the maximum principle once again, we obtain

$$|h(z)(1-z^2)^2| \le 8e^A \qquad (z \in \mathbb{D},\ \operatorname{Im} z \ge 0).$$

This completes the proof of the lemma. □

Proof of Theorem 9.6.4 Let $\phi \in \operatorname{Hol}_{\mathcal{B}_e,\mathcal{N}^+}(\mathbb{C}_\infty \setminus E)$.

Since $\phi|\mathcal{D}_e \in \mathcal{B}_e$, we can write $\phi(z) = \sum_{k\ge0} b_k/z^{k+1}$ on $|z|>1$, where $\sum_{k\ge0} |b_k|^2/(k+1) = \|\phi\|^2_{\mathcal{B}_e}$. By the Cauchy–Schwarz inequality, it follows that

$$|\phi(z)| \le \Bigl(\sum_{k\ge0}\frac{|b_k|^2}{k+1}\Bigr)^{1/2}\Bigl(\sum_{k\ge0}\frac{k+1}{|z|^{2k+2}}\Bigr)^{1/2} = \frac{\|\phi\|_{\mathcal{B}_e}}{|z|-1} \qquad (|z|>1).$$

Also, since $\phi|\mathbb{D} \in \mathcal{N}^+$, the radial limit $\phi^*(\zeta) := \lim_{r\to1^-}\phi(r\zeta)$ exists a.e. on $\mathbb{T}$, the function $\log|\phi^*|$ is integrable on $\mathbb{T}$, and

$$\log|\phi(z)| \le \frac{1}{2\pi}\int_{\mathbb{T}} \operatorname{Re}\Bigl(\frac{\zeta+z}{\zeta-z}\Bigr)\log|\phi^*(\zeta)|\,|d\zeta| \le \frac{2\bigl\|\log|\phi^*|\bigr\|_{L^1(\mathbb{T})}}{1-|z|} \qquad (|z|<1).$$

Combining these bounds, we see that there are constants C_1, C_2 such that

$$|\phi(z)| \le \begin{cases} C_1/(|z|-1), & |z|>1 \\ \exp(C_2/(1-|z|)), & |z|<1. \end{cases} \tag{9.17}$$

Now let Δ be an open disk of radius $r<1$ whose center ζ lies on $\mathbb{T}$, and such that $\Delta \cap E = \emptyset$. Let m be the Möbius transformation that maps $\mathbb{D}$ onto Δ, maps 0 to ζ, and maps the upper half-plane onto the exterior of the unit disk. Set $h := \phi \circ m$. Then the estimates (9.17) for ϕ translate in bounds of the type (9.16) for h. Lemma 9.6.5 therefore applies and, after translating back again, we obtain a bound of the form $|\phi(z)| \le C'/\operatorname{dist}(z, \mathbb{T} \setminus \Delta)^2$ for $z \in \mathbb{D}_e \cap \Delta$. Moreover, an inspection of the proof shows that the constant C' can be chosen independently of Δ, so $|\phi(z)| \le C'/\operatorname{dist}(z, E)^2$ for $z \in \mathbb{D}_e \cap U$, where U is the union of all such disks Δ.

It remains to take care of the case $z \in \mathbb{D}_e \setminus U$. For these z, it is easy to see that $\operatorname{dist}(z, E) \asymp |z| - 1$. Therefore, the first bound in (9.17) already implies the existence of a constant C'' such that $|\phi(z)| \le C''/\operatorname{dist}(z, E)$ $(z \in \mathbb{D}_e \setminus U)$. If we further restrict z to satisfy $1 < |z| < 2$, then $|\phi(z)| \le 3C''/\operatorname{dist}(z, E)^2$.

Finally taking $C := \max\{C', 3C''\}$, we obtain (9.15). □

Theorem 9.6.6 *Let E be a closed subset of $\mathbb{T}$, and let $f \in \mathcal{D}$ be a function satisfying $|f(z)| \le C \operatorname{dist}(z, E)^2$ on $\mathbb{D}$. Then*

$$\operatorname{Hol}_{\mathcal{B}_e, \mathcal{N}^+}(\mathbb{C}_\infty \setminus E) \subset [f]^\perp_{\mathcal{D}}. \tag{9.18}$$

Proof Let $\phi \in \operatorname{Hol}_{\mathcal{B}_e, \mathcal{N}^+}(\mathbb{C}_\infty \setminus E)$. If we write $f(z) = \sum_{k \ge 0} a_k z^k$ and $\phi(z) = \sum_{k \ge 0} b_k / z^{k+1}$, then

$$\langle f, \phi \rangle = \sum_{k \ge 0} a_k b_k = \lim_{r \to 1^-} \sum_{k \ge 0} a_k b_k r^{2k+1} = \lim_{r \to 1^-} \frac{1}{2\pi i} \int_{\mathbb{T}} f(r\zeta) \phi(\zeta / r)\, d\zeta.$$

By hypothesis $|f(z)| \le C \operatorname{dist}(z, E)^2$ on $\mathbb{D}$, and by Theorem 9.6.4 we have $|\phi(z)| \le C'/\operatorname{dist}(z, E)^2$ on $1 < |z| < 2$. Consequently the family of functions $\zeta \mapsto f(r\zeta)\phi(\zeta/r)$ $(1/2 < r < 1)$ is uniformly bounded on $\mathbb{T}$, and we can pass the limit inside the integral above to obtain

$$\langle f, \phi \rangle = \frac{1}{2\pi i} \int_{\mathbb{T}} f^*(\zeta) \phi(\zeta)\, d\zeta. \tag{9.19}$$

(Note that the assumption on f forces E to be of measure zero, so ϕ is continuous a.e. on $\mathbb{T}$.)

Now $\phi|_{\mathbb{D}} \in \mathcal{N}^+$ and $f \in \mathcal{D}$, so also $f\phi \in \mathcal{N}^+$. Further, as we have just seen, $f^*\phi \in L^\infty(\mathbb{T})$. By Smirnov's maximum principle (see Theorem A.3.8 in Appendix A), it follows that $f\phi \in H^\infty$. This implies that the integral in (9.19) vanishes, and consequently $\langle f, \phi \rangle = 0$.

The same argument shows that $\langle pf, \phi \rangle = 0$ for all polynomials p. We conclude that $\phi \in [f]^\perp_{\mathcal{D}}$. □

We are now (at last!) in a position to produce some examples of non-empty Bergman–Smirnov exceptional sets.

Theorem 9.6.7 *Every finite subset of $\mathbb{T}$ is Bergman–Smirnov exceptional.*

Proof Given a finite subset E of $\mathbb{T}$, let f be the polynomial with a double root at each point of E. Theorem 9.6.6 applies, and so $\mathrm{Hol}_{\mathcal{B}_e,\mathcal{N}^+}(\mathbb{C}_\infty \setminus E) \subset [f]_{\mathcal{D}}^{\perp}$. But also, from Theorem 9.2.1 we know that f is cyclic for $\mathcal{D}$, so $[f]_{\mathcal{D}}^{\perp} = \{0\}$. Thus $\mathrm{Hol}_{\mathcal{B}_e,\mathcal{N}^+}(\mathbb{C}_\infty \setminus E) = \{0\}$, and E is Bergman–Smirnov exceptional. □

Theorem 9.6.8 *Let E be a closed subset of $\mathbb{T}$ such that*

$$\int_0^{1/e} c(E_t) \frac{\log\log(1/t)}{t\log(1/t)}\,dt < \infty.$$

Then E is a Bergman–Smirnov exceptional set.

Proof Just as in the proof of Theorem 9.3.5, the hypothesis on E allows us to construct a holomorphic family of functions $(g_\lambda)_{\lambda\in\Omega}$ in $\mathcal{D}$, where Ω is the sector $|\arg(\lambda)| < \pi/6$, and $|g_\lambda(z)| \le C_\lambda \operatorname{dist}(z, E)^{|\lambda|/2}$ and $\lim_{\lambda\to 0} \|g_\lambda - 1\|_{\mathcal{D}} = 0$.

Let $\phi \in \mathrm{Hol}_{\mathcal{B}_e,\mathcal{N}^+}(\mathbb{C}_\infty \setminus E)$. If $|\lambda| > 4$, then $|g_\lambda(z)| \le C \operatorname{dist}(z, E)^2$, and Theorem 9.6.6 then shows that $\langle pg_\lambda, \phi\rangle = 0$ for all polynomials p. By the identity principle, it follows that $\langle pg_\lambda, \phi\rangle = 0$ for all $\lambda \in \Omega$ and all p. Letting $\lambda \to 0$, it follows that $\langle p, \phi\rangle = 0$ for all p. As polynomials are dense in $\mathcal{D}$, we deduce that $\phi = 0$. Thus E is Bergman–Smirnov exceptional. □

Theorem 9.6.9 *Let E be a closed subset of $\mathbb{T}$ such that, for some $\sigma > 0$,*

$$|E_t| = O(t^\sigma) \quad (t \to 0) \qquad \text{and} \qquad \int_0^\pi \frac{dt}{|E_t|} = \infty.$$

Then E is a Bergman–Smirnov exceptional set.

Proof Just as in the proof of Theorem 9.4.1, the assumptions on E allow us to construct a cyclic function g for $\mathcal{D}$ such that $|g(z)| \le C \operatorname{dist}(z, E)^\epsilon$ on $\mathbb{D}$, where $\epsilon > 0$. Replacing g by a high enough power g^n, we may further suppose that $\epsilon = 2$. By Theorem 9.6.6, we then have $\mathrm{Hol}_{\mathcal{B}_e,\mathcal{N}^+}(\mathbb{C}_\infty \setminus E) \subset [g]_{\mathcal{D}}^{\perp} = \{0\}$. Therefore E is Bergman–Smirnov exceptional. □

The next result extends our library of Bergman–Smirnov exceptional sets still further, using a quite different technique.

Theorem 9.6.10 *Let E be a closed subset of $\mathbb{T}$. Suppose that E contains a Bergman–Smirnov exceptional set E' such that $E \setminus E'$ is countable. Then E is itself Bergman–Smirnov exceptional.*

The following special case is particularly worthy of note.

Corollary 9.6.11 *Every countable closed subset of $\mathbb{T}$ is a Bergman–Smirnov exceptional set.* □

The proof of Theorem 9.6.10 uses the following decomposition lemma.

Lemma 9.6.12 *Let E_1, E_2 be disjoint closed subsets of $\mathbb{T}$, and $E := E_1 \cup E_2$. Then every $\phi \in \mathrm{Hol}_{\mathcal{B}_e,\mathcal{N}^+}(\mathbb{C}_\infty \setminus E)$ can be decomposed as $\phi = \phi_1 + \phi_2$, where $\phi_j \in \mathrm{Hol}_{\mathcal{B}_e,\mathcal{N}^+}(\mathbb{C}_\infty \setminus E_j)$ for $j = 1, 2$.*

Proof Using Cauchy's integral formula in the standard way, we can decompose ϕ as $\phi = \phi_1 + \phi_2$, where $\phi_j \in \mathrm{Hol}(\mathbb{C}_\infty \setminus E_j)$ $(j = 1, 2)$. Adding and subtracting a constant, we may also ensure that $\phi_1(\infty) = \phi_2(\infty) = 0$. The point is to show that $\phi_j|_{\mathbb{D}_e} \in \mathcal{B}_e$ and that $\phi_j|_{\mathbb{D}} \in \mathcal{N}^+$.

We begin with $\phi_j|_{\mathbb{D}_e}$. Let U_1, U_2 be bounded disjoint open neighborhoods of E_1, E_2 respectively. Then

$$\|\phi_1\|_{\mathcal{B}_e}^2 = \frac{1}{\pi}\int_{\mathbb{D}_e \cap U_1} |\phi_1(z)|^2 \frac{dA(z)}{|z|^2} + \frac{1}{\pi}\int_{\mathbb{D}_e \setminus U_1} |\phi_1(z)|^2 \frac{dA(z)}{|z|^2}.$$

The second integral is finite, because ϕ_1 is bounded on $\mathbb{D}_e \setminus U_1$ and $\phi_1(z) = O(1/z)$ as $|z| \to \infty$. The first integral is finite because $\phi_1 = \phi - \phi_2$, where $\phi|_{\mathbb{D}_e} \in \mathcal{B}_e$ and ϕ_2 is bounded on U_1. It follows that $\phi_1|_{\mathbb{D}_e} \in \mathcal{B}_e$. Likewise for ϕ_2.

Now we consider $\phi_j|_{\mathbb{D}}$. To show that $\phi_j|_{\mathbb{D}} \in \mathcal{N}^+$, we use the characterization of $\mathcal{N}^+$ as the set of $h \in \mathrm{Hol}(\mathbb{D})$ such that $\{\zeta \mapsto \log^+ |h(r\zeta)| : 0 < r < 1\}$ is uniformly integrable on $\mathbb{T}$ (see Theorem A.3.7 in Appendix A). Once again, let U_1, U_2 be disjoint open neighborhoods of E_1, E_2. On $\mathbb{T} \setminus U_1$, the family $\{\log^+ |\phi_1(r\zeta)| : 0 < r < 1\}$ is uniformly bounded, so it is certainly uniformly integrable. On $\mathbb{T} \cap U_1$,

$$\log^+ |\phi_1(r\zeta)| = \log^+ |\phi(r\zeta) - \phi_2(r\zeta)| \le \log^+ |\phi(r\zeta)| + \log^+ |\phi_2(r\zeta)| + \log 2,$$

where the right-hand side is the sum of a uniformly integrable family (because $\phi|_{\mathbb{D}} \in \mathcal{N}^+$) and a uniformly bounded family. Putting this all together we deduce that $\{\log^+ |\phi_1(r\zeta)| : 0 < r < 1\}$ is uniformly integrable on $\mathbb{T}$, and hence that $\phi_1|_{\mathbb{D}} \in \mathcal{N}^+$. The case of ϕ_2 is similar. □

Proof of Theorem 9.6.10 Let $\phi \in \mathrm{Hol}_{\mathcal{B}_e,\mathcal{N}^+}(\mathbb{C}_\infty \setminus E)$. Let F be the closed subset of E consisting of those points across which ϕ has no holomorphic continuation. Clearly we then have $\phi \in \mathrm{Hol}_{\mathcal{B}_e,\mathcal{N}^+}(\mathbb{C}_\infty \setminus F)$.

We claim that $F \subset E'$. Indeed, let us suppose the contrary. Then $F \setminus E'$ is non-empty, countable and closed in $\mathbb{T} \setminus E'$. By the Baire category theorem, F contains an isolated point ζ. Set $F_1 := \{\zeta\}$ and $F_2 := F \setminus \{\zeta\}$. By Lemma 9.6.12, we can write $\phi = \phi_1 + \phi_2$, where $\phi_j \in \mathrm{Hol}_{\mathcal{B}_e,\mathcal{N}^+}(\mathbb{C}_\infty \setminus F_j)$. As F_1 is a singleton, it is Bergman–Smirnov exceptional, and so $\phi_1 = 0$. Consequently $\phi = \phi_2$, which is holomorphic at ζ, contradicting the definition of F. The claim is justified.

Since $F \subset E'$ and E' is Bergman–Smirnov exceptional, so too is F, and therefore, finally, $\phi = 0$, as required. □

All these examples tend to suggest that the converse of Theorem 9.6.3 is true. Hedenmalm and Shields [58] posed this formally as a problem.

Problem 9.6.13 (Hedenmalm–Shields) *Let E be a closed subset of $\mathbb{T}$ such that $c(E) = 0$. Must E be a Bergman–Smirnov exceptional set?*

This problem is still open. Note that, thanks to Theorem 9.6.2, an affirmative answer would imply the Brown–Shields conjecture, at least for those f continuous up to the boundary.

We conclude by remarking that, if we replace $\mathcal{N}^+$ by $\mathcal{B}$, the usual Bergman space on $\mathbb{D}$, then $\mathrm{Hol}_{\mathcal{B}_e,\mathcal{B}}(\mathbb{C}_\infty \setminus E) = \{0\}$ precisely when E is of logarithmic capacity zero. A proof is sketched in Exercise 9.6.3 below. This rules out the possibility of answering Problem 9.6.13 negatively by using the Cauchy transform of a measure on $\mathbb{T}$, as in the proof of Theorem 9.6.3.

Exercises 9.6

1. Let $f \in \mathcal{D} \cap A(\mathbb{D})$ be an outer function and let $E := \{\zeta \in \mathbb{T} : f(\zeta) = 0\}$. Let $g \in \mathcal{D}$ and suppose that $|g(z)| \le C\,\mathrm{dist}(z, E)^2$ on $\mathbb{D}$. Show that
$$[f]_{\mathcal{D}}^{\perp} \subset \mathrm{Hol}_{\mathcal{B}_e,\mathcal{N}^+}(\mathbb{C}_\infty \setminus E) \subset [g]_{\mathcal{D}}^{\perp},$$
and deduce that $g \in [f]_{\mathcal{D}}$. (Compare Theorem 9.3.3.)
2. Show that the union of two Bergman–Smirnov exceptional sets is again a Bergman–Smirnov exceptional set.
3. Let E be a closed subset of $\mathbb{T}$ of logarithmic capacity zero, and let $f \in \mathrm{Hol}_{\mathcal{B}_e,\mathcal{B}}(\mathbb{C} \setminus E)$.
   (i) Let $s > 0$ and set $E_s := \{\zeta \in \mathbb{T} : d(\zeta, E) \le s\}$. Let $K(t) = \log^+(2/t)$ and let ν_s be the equilibrium measure for E_s. Prove that
   $$\lim_{r\to 1^-} \frac{1}{2\pi}\int_{\mathbb{T}} \Big(K\nu_s(\zeta) - \frac{1}{c(E_s)}\Big)\Big(f(r\zeta) - f(\zeta/r)\Big)\,|d\zeta| = 0.$$
   (ii) Deduce that
   $$\lim_{r\to 1^-} \frac{1}{2\pi}\int_{\mathbb{T}} K\nu_s(\zeta)\Big(f(r\zeta) - f(\zeta/r)\Big)\,|d\zeta| = \frac{f(0)}{c(E_s)}.$$
   (iii) Using the result of Theorem 2.4.4, prove that, for $0 < r < 1$:
   $$\Big|\frac{1}{2\pi}\int_{\mathbb{T}} K\nu_s(\zeta) f(\zeta/r)\,|d\zeta|\Big| \le \big\|f|_{\mathbb{D}_e}\big\|_{\mathcal{B}_e} c(E_s)^{-1/2},$$
   $$\Big|\frac{1}{2\pi}\int_{\mathbb{T}} K\nu_s(\zeta) f(r\zeta)\,|d\zeta|\Big| \le 2\big\|f|_{\mathbb{D}}\big\|_{\mathcal{B}} c(E_s)^{-1/2} + |f(0)|/c(\mathbb{T}).$$
   (iv) By combining (ii) and (iii) and letting $s \to 0$, deduce that $f(0) = 0$.
   (v) Repeating with f replaced by $z^n f$ ($n \in \mathbb{Z}$), conclude that $f \equiv 0$.

Notes on Chapter 9

§9.1

The study of cyclicity in $\mathcal{D}$ was instigated by Brown and Shields in [24]. Their results were subsequently improved and extended to $\mathcal{D}_\mu$ by Richter and Sundberg, whose article [99] is the source for the results in this section. Exercise 9.1.2 is based on a result in [13].

§9.2

Theorem 9.2.3 goes back to Carleson [26]. The proof given here is from [24], where it is attributed to J. Shapiro. Theorem 9.2.7 is due to Brown and Cohn [23]. Exercise 9.2.1 is based on a result in [13]. The first four parts of Exercise 9.2.3 are taken from [99]. Part (v) is a result of Beurling [16].

§9.3

As mentioned in the text, Conjecture 9.3.2 is due to Brown and Shields [24]. The proof of Theorem 9.3.3 is from [44], adapted from a technique of Korenblum [67]. Theorem 9.3.5 is from [43]. Exercise 9.3.2 is based on a suggestion of Richter (private communication).

§9.4

Theorem 9.4.3 can be made more precise as follows. Let us write $n_E(t) := 2\sum_j 1_{\{|I_j|>2t\}}$ where (I_j) are the components of $\mathbb{T} \setminus E$. Then, under the same hypotheses as in the theorem, we have

$$\mathcal{D}(f_{w,E}) \asymp \int_0^\pi w'(t)^2 t n_E(t)\, dt,$$

where the implied constants depend only on γ. There are also versions of this result for the case $\gamma \leq 2$, as well as for when w is a decreasing function. For more on distance functions, we refer to the article [44], on which this section is based.

§9.5

The principal results in this section are due to Hedenmalm and Shields [58]. Using a somewhat different technique, Richter and Sundberg [100] showed that these results can be generalized to functions not necessarily belonging to the disk algebra, provided that one defines $E := \{\zeta : \liminf_{z\to\zeta} |f(z)| = 0\}$.

§9.6

The notion of Bergman–Smirnov exceptional set was introduced by Hedenmalm and Shields in [58]. They proved that such sets are of logarithmic capacity zero, and posed the question as to whether the converse is true. They also showed that every countable compact subset of $\mathbb{T}$ is Bergman–Smirnov exceptional. The first uncountable examples were given in [43]. Theorem 9.6.8 as well as the results used in its proof are taken from this source. Lemma 9.6.5 is a result of Taylor and Williams [118, Lemma 5.8], formulated here in different terms, and with a slightly different proof. Exercise 9.6.3 is a result of Kahane and Salem [61].

Appendix A

Hardy spaces

This appendix is intended as a brief summary of the basic elements of the theory of Hardy spaces. For the most part, we state results without proof. Unless explicit references are given, proofs can be found in [104, Chapter 17], which provides a succinct introduction to the subject. The reader seeking more details is referred to the specialized texts on Hardy spaces [40, 50, 66, 77].

A.1 Hardy spaces

Definition A.1.1 For $0 < p < \infty$, the *Hardy space* H^p is defined by

$$H^p := \Big\{ f \in \mathrm{Hol}(\mathbb{D}) : \sup_{0<r<1} \int_0^{2\pi} |f(re^{i\theta})|^p \, d\theta < \infty \Big\}.$$

Also we define

$$H^\infty := \Big\{ f \in \mathrm{Hol}(\mathbb{D}) : \sup_{z \in \mathbb{D}} |f(z)| < \infty \Big\}.$$

Finally, the *Nevanlinna class* $\mathcal{N}$ is defined by

$$\mathcal{N} := \Big\{ f \in \mathrm{Hol}(\mathbb{D}) : \sup_{0<r<1} \int_0^{2\pi} \log^+ |f(re^{i\theta})| \, d\theta < \infty \Big\}.$$

All the Hardy spaces H^p are vector spaces, and the Nevanlinna class $\mathcal{N}$ is even an algebra. Also $H^\infty \subset H^{p_2} \subset H^{p_1} \subset \mathcal{N}$ whenever $0 < p_1 < p_2 < \infty$.

Theorem A.1.2 *If $1 \le p < \infty$, then H^p is a Banach space with respect to the norm*

$$\|f\|_{H^p} := \sup_{0<r<1} \Big(\frac{1}{2\pi} \int_0^{2\pi} |f(re^{i\theta})|^p \, d\theta \Big)^{1/p}.$$

Also H^∞ is a Banach algebra with respect to the norm $\|f\|_{H^\infty} := \sup_{z \in \mathbb{D}} |f(z)|$.

Because of its central role in the theory, H^2 is often called *the* Hardy space.

Theorem A.1.3 *H^2 is a Hilbert space and, writing $f(z) = \sum_{k\geq 0} a_k z^k$, we have*

$$\|f\|_{H^2}^2 = \sum_{k\geq 0} |a_k|^2.$$

Given $f \in \text{Hol}(\mathbb{D})$ and $\zeta \in \mathbb{T}$, we write $f^*(\zeta) := \lim_{r\to 1^-} f(r\zeta)$ whenever this radial limit exists.

Theorem A.1.4 (Fatou's theorem) *If $f \in \mathcal{N}$, then f^* exists a.e. on $\mathbb{T}$ in the sense of Lebesgue measure. Moreover, for a.e. $\zeta \in \mathbb{T}$, we have $f(z) \to f^*(\zeta)$ as $z \to \zeta$ in each non-tangential approach region $|z - \zeta| \leq \kappa(1 - |z|)$ $(\kappa > 0)$.*

Theorem A.1.5 *Let $1 \leq p \leq \infty$. If $f \in H^p$, then $f^* \in L^p(\mathbb{T})$, and we have $\|f\|_{H^p} = \|f^*\|_{L^p(\mathbb{T})}$. Furthermore, f is equal to the Cauchy and Poisson integrals of f^*, namely*

$$f(z) = \frac{1}{2\pi i}\int_{\mathbb{T}} \frac{f^*(\zeta)}{\zeta - z}\,d\zeta \qquad (z \in \mathbb{D})$$

and

$$f(z) = \frac{1}{2\pi}\int_{\mathbb{T}} \frac{1 - |z|^2}{|\zeta - z|^2} f^*(\zeta)\,|d\zeta| \qquad (z \in \mathbb{D}).$$

Theorem A.1.6 *If $f \in \mathcal{N}$ and $f \not\equiv 0$, then $\log|f^*| \in L^1(\mathbb{T})$.*

Corollary A.1.7 *If $f \in \mathcal{N}$ and $f^* = 0$ on a subset of $\mathbb{T}$ of positive Lebesgue measure, then $f \equiv 0$.*

Theorem A.1.8 *Let $f \in \mathcal{N}$ with $f \not\equiv 0$, and let (z_n) be the zeros of f, listed according to multiplicity. Then*

$$\sum_n (1 - |z_n|) < \infty.$$

Theorem A.1.9 (Hardy's inequality) *Let $f \in H^1$, say $f(z) := \sum_{k\geq 0} a_k z^k$. Then*

$$\sum_{k\geq 0} \frac{|a_k|}{k+1} \leq \pi \|f\|_{H^1}.$$

Proof See [40, Corollary to Theorem 3.15]. □

We conclude with two results concerning H^p for $p < 1$.

Theorem A.1.10 *If $f \in \text{Hol}(\mathbb{D})$ and f is univalent, then we have $f \in H^p$ for all $p < 1/2$.*

Proof See [40, Theorem 3.16] or for a shorter proof [91, Theorem 5.1]. □

The example of the Koebe function $f(z) := z/(1-z)^2$, which is univalent but does not belong $H^{1/2}$, shows that this theorem is sharp.

Theorem A.1.11 *If $f \in \mathrm{Hol}(\mathbb{D})$ and $\operatorname{Re} f(z) > 0$ for all $z \in \mathbb{D}$, then $f \in H^p$ for all $p < 1$.*

Proof Fix $p \in (0,1)$. Since the principal branch of $w \mapsto w^p$ maps the half-plane $\{\operatorname{Re} w > 0\}$ into the sector $\{|\arg w| < p\pi/2\}$, the function f satisfies $|f(z)|^p \le \operatorname{Re}(f(z)^p)/\cos(p\pi/2)$ for all $z \in \mathbb{D}$. Consequently

$$\frac{1}{2\pi}\int_0^{2\pi} |f(re^{i\theta})|^p \, d\theta \le \frac{1}{2\pi}\int_0^{2\pi} \frac{\operatorname{Re}(f(re^{i\theta}))^p}{\cos(p\pi/2)}\, d\theta = \frac{\operatorname{Re}(f(0)^p)}{\cos(p\pi/2)} \qquad (0<r<1),$$

the last equality because $\operatorname{Re}(f^p)$ is a harmonic function. As the final bound is independent of r, we conclude that $f \in H^p$. □

The function $f(z) := 1/(1-z)$ $(z \in \mathbb{D})$, which has positive real part but does not belong to H^1, shows that this result is sharp.

There is a useful corollary concerning the Cauchy transforms of measures. Given a complex measure μ on $\mathbb{T}$, we define its *Cauchy transform* by

$$C\mu(z) := \int_{\mathbb{T}} \frac{d\mu(\zeta)}{\zeta - z} \qquad (z \in \mathbb{D}).$$

Corollary A.1.12 *If μ is a complex measure on $\mathbb{T}$, then $C\mu \in H^p$ for all $0 < p < 1$.*

Proof It suffices to observe that $C\mu$ can be written as a linear combination of holomorphic functions with positive real parts. Indeed,

$$C\mu(z) = \int_{\mathbb{T}} \frac{\overline{\zeta}\, d\mu(\zeta)}{1-\overline{\zeta}z} = \sum_{j=1}^{4} i^j \int_{\mathbb{T}} \frac{d\mu_j(\zeta)}{1-\overline{\zeta}z} \qquad (z \in \mathbb{D}),$$

where μ_j are positive measures, and

$$\operatorname{Re}\Big(\int_{\mathbb{T}} \frac{d\mu_j(\zeta)}{1-\overline{\zeta}z}\Big) = \int_{\mathbb{T}} \operatorname{Re}\Big(\frac{1}{1-\overline{\zeta}z}\Big)\, d\mu_j(\zeta) \ge 0 \qquad (z \in \mathbb{D}).$$ □

The example $\mu := \delta_1$, for which $C\mu(z) = (1-z)^{-1}$, shows that we may have $C\mu \notin H^1$.

A.2 Inner and outer functions

Definition A.2.1 An *inner function* is a function $f \in H^\infty$ such that $|f^*| = 1$ a.e. on $\mathbb{T}$.

Here are two classes of examples of inner functions.

Definition A.2.2 Let (z_n) be a (finite or infinite) sequence in $\mathbb{D}$ satisfying $\sum_n(1 - |z_n|) < \infty$. The *Blaschke product* with zeros (z_n) is defined by

$$B(z) := \prod_n \frac{|z_n|}{z_n}\Big(\frac{z_n - z}{1 - \overline{z}_n z}\Big) \qquad (z \in \mathbb{D}).$$

Definition A.2.3 Let σ be a finite positive Borel measure on $\mathbb{T}$ that is singular with respect to Lebesgue measure. The *singular inner function* corresponding to σ is defined by

$$S(z) := \exp\Big(-\int_{\mathbb{T}} \frac{\zeta + z}{\zeta - z}\, d\sigma(\zeta)\Big) \qquad (z \in \mathbb{D}).$$

Theorem A.2.4 *A function f is inner if and only if $f = cBS$, where B is a Blaschke product, S is a singular inner function, and c is a unimodular constant.*

Definition A.2.5 An *outer function* is a function of the form

$$F(z) = \exp\Big(\frac{1}{2\pi}\int_{\mathbb{T}} \frac{\zeta + z}{\zeta - z} \log \phi(\zeta)\, |d\zeta|\Big) \qquad (z \in \mathbb{D}), \qquad \text{(A.1)}$$

where $\phi : \mathbb{T} \to [0, \infty)$ is a function such that $\log \phi \in L^1(\mathbb{T})$.

Theorem A.2.6 *Let F be the outer function given by* (A.1). *Then $F \in \mathcal{N}$ and $|F^*| = \phi$ a.e. on $\mathbb{T}$. Moreover, for each $p \in (0, \infty]$, we have $F \in H^p$ if and only if $\phi \in L^p(\mathbb{T})$.*

Theorem A.2.7 (Canonical factorization) *Let $p \in (0, \infty]$. If $f \in H^p$ with $f \not\equiv 0$, then $f = cBSF$, where B is a Blaschke product, S is a singular inner function, F is an outer function in H^p, and c is a unimodular constant. This factorization is unique.*

Corollary A.2.8 (Inner-outer factorization) *Let $p \in (0, \infty]$. If $f \in H^p$ with $f \not\equiv 0$, then $f = f_i f_o$ where f_i is inner and f_o is outer with $f_o \in H^p$. This factorization is unique.*

There is an important connection between inner-outer factorization and the closed invariant subspaces of the shift operator $M_z : f \mapsto zf$ on H^2. We write $\mathrm{Lat}(M_z, H^2)$ for the family of all such subspaces.

Theorem A.2.9 (Beurling) *$\mathcal{M} \in \mathrm{Lat}(M_z, H^2) \setminus \{0\}$ if and only if $\mathcal{M} = \theta H^2$, where θ is an inner function. Moreover θ is unique up to multiplication by a unimodular constant.*

We say that $f \in H^2$ is *cyclic* for M_z if it is contained in no proper closed shift-invariant subspace.

Corollary A.2.10 *Let $f \in H^2$. Then f is cyclic if and only if it is outer.*

A.3 The Smirnov class

For the Nevanlinna class $\mathcal{N}$, the canonical factorization theorem holds in a modified form.

Theorem A.3.1 *If $f \in \mathcal{N}$ with $f \not\equiv 0$, then $f = cB(S_1/S_2)F$, where B is a Blaschke product, S_1, S_2 are singular inner functions, F is an outer function and c is a unimodular constant.*

Corollary A.3.2 *$f \in \mathcal{N}$ if and only if $f = g/h$, where $g, h \in H^\infty$ and $h(z) \neq 0$ for all $z \in \mathbb{D}$.*

The functions $f \in \mathcal{N}$ for which Theorem A.3.1 holds with $S_2 = 1$ form a proper subclass, of interest in its own right.

Definition A.3.3 The *Smirnov class* $\mathcal{N}^+$ consists of all functions of the form $f = f_i f_o$, where f_i is inner and f_o is outer, together with the zero function.

We have the inclusions $H^p \subset \mathcal{N}^+ \subset \mathcal{N}$ for all $p > 0$. They are all proper.

There are several alternative characterizations of the Smirnov class. The first of these follows easily from the definition.

Theorem A.3.4 *$f \in \mathcal{N}^+$ if and only if $f = g/h$, where $g, h \in H^\infty$ and h is outer.*

Corollary A.3.5 *$\mathcal{N}^+$ is a subalgebra of $\mathcal{N}$.*

Theorem A.3.6 *Let $f \in \mathcal{N}$. Then $f \in \mathcal{N}^+$ if and only if*

$$\lim_{r \to 1^-} \int_0^{2\pi} \log^+ |f(re^{i\theta})| \, d\theta = \int_0^{2\pi} \log^+ |f^*(e^{i\theta})| \, d\theta. \tag{A.2}$$

Proof See for example [40, Theorem 2.10]. □

Theorem A.3.7 *Let $f \in$ Hol$(\mathbb{D})$. Write $f_r(z) := f(rz)$ $(0 < r < 1)$. Then $f \in \mathcal{N}^+$ if and only if the family $\{\log^+ |f_r|\}_{0<r<1}$ is uniformly integrable on $\mathbb{T}$.*

Proof Suppose that $\{\log^+ |f_r|\}_{0<r<1}$ is uniformly integrable. Uniformly integrable implies L^1-bounded, so certainly $f \in \mathcal{N}$. It follows that the radial limit

f^* exists a.e. on $\mathbb{T}$, and so $\log^+|f_r| \to \log^+|f^*|$ a.e. as $r \to 1^-$. Uniform integrability then implies that $\log^+|f_r| \to \log^+|f^*|$ in $L^1(\mathbb{T})$, which gives (A.2). Hence $f \in \mathcal{N}^+$.

Conversely, suppose that $f \in \mathcal{N}^+$ and $f \not\equiv 0$. Let f_o be its outer factor. Then $\log|f_o^*| \in L^1(\mathbb{T})$, and $\log|f_o|$ is its Poisson integral, so we have $\log|(f_o)_r| \to \log|f_o^*|$ in $L^1(\mathbb{T})$. It follows that $\{\log|(f_o)_r|\}_{0<r<1}$ is a uniformly integrable family. Since $|f| \le |f_o|$ on $\mathbb{D}$, we have $\log^+|f_r| \le \big|\log|(f_o)_r|\big|$, and we conclude that $\{\log^+|f_r|\}_{0<r<1}$ is uniformly integrable. □

Finally, we need the following result, which follows fairly easily from the definition of $\mathcal{N}^+$.

Theorem A.3.8 (Smirnov's maximum principle) *Let $p \in (0, \infty]$. If $f \in \mathcal{N}^+$ and $f^* \in L^p(\mathbb{T})$, then $f \in H^p$.*

Proof See for example [40, Theorem 2.11]. □

Appendix B

The Hardy–Littlewood maximal function

B.1 Weak-type inequality for the maximal function

Definition B.1.1 Given $h \in L^1(\mathbb{T})$, its *Hardy–Littlewood maximal function* $Mh : \mathbb{T} \to [0, \infty]$ is defined by

$$(Mh)(\zeta) := \sup_{\delta>0} \frac{1}{2\delta} \int_{|\theta - \arg \zeta| < \delta} |h(e^{i\theta})| \, d\theta \qquad (\zeta \in \mathbb{T}).$$

Clearly Mh is lower semicontinuous. It also satisfies the following weak-type inequality.

Theorem B.1.2 *If* $h \in L^1(\mathbb{T})$, *then*

$$\Big|\{\zeta \in \mathbb{T} : Mh(\zeta) > t\}\Big| \leq 6\pi \|h\|_1 / t \qquad (t > 0).$$

Proof Fix $t > 0$ and let K be a compact subset of $\{Mh > t\}$. For each $\zeta \in K$, there exists an open arc I_ζ centered at ζ such that

$$\frac{1}{|I_\zeta|} \int_{I_\zeta} |h| \, d\theta > t.$$

The collection $\{I_\zeta : \zeta \in K\}$ forms an open cover of K, so there exists a finite subcover, $I_{\zeta_1}, \ldots, I_{\zeta_m}$ say. Choose a member J_1 of this subcover with $|J_1|$ as large as possible. Choose a member J_2 of this subcover disjoint from J_1 with $|J_2|$ as large as possible. Choose a member J_3 of this subcover disjoint from $J_1 \cup J_2$ with $|J_3|$ as large as possible. And so on. Eventually the process stops. Let $J_1, \ldots, J_n$ be the totality of the arcs thereby obtained.

For each ζ_j, there exists k such that $I_{\zeta_j} \cap J_k \neq \emptyset$ and $|J_k| \geq |I_{\zeta_j}|$. This implies that $I_{\zeta_j} \subset 3J_k$, where $3J_k$ denotes the arc with the same center as J_k and three times the length. Thus the family $\{3J_k : k = 1, \ldots, n\}$ is a cover of K, and consequently

$$|K| \leq \sum_{k=1}^{n} 3|J_k| \leq \sum_{k=1}^{n} \frac{3}{t} \int_{J_k} |h| \, d\theta = \frac{3}{t} \int_{\cup_k J_k} |h| \, d\theta \leq \frac{6\pi \|h\|_1}{t}.$$

As K is an arbitrary compact subset of the open set $\{Mh > t\}$, the result follows.

□

Appendix C

Positive definite matrices

C.1 Basic facts about positive definite matrices

Definition C.1.1 A complex $n \times n$ matrix $A = (a_{jk})$ is called *positive semi-definite* if, for any choice of complex numbers $\lambda_1, \dots, \lambda_n$, we have

$$\sum_{j=1}^{n} \sum_{k=1}^{n} a_{jk} \lambda_j \overline{\lambda}_k \geq 0. \tag{C.1}$$

It is *positive definite* if, further, equality holds only when all the λ_j are zero.

It is easy to check that, if condition (C.1) holds, then $a_{jk} = \overline{a}_{kj}$ for all j, k Thus, if A is positive semi-definite, then it is automatically a hermitian matrix.

Now let A be an $n \times n$ hermitian matrix. According to the spectral theorem, it has n real eigenvalues $\sigma_1, \dots, \sigma_n$ and n orthonormal eigenvectors $u_1, \dots, u_n$. Hence, we have the representation

$$A = U \Sigma U^*, \tag{C.2}$$

where Σ is the diagonal matrix whose entries are $\sigma_1, \dots, \sigma_n$ and U is the unitary matrix whose columns are $u_1, \dots, u_n$. According to this decomposition, A is positive semi-definite if and only if $\sigma_i \geq 0$ for all i. Similarly, A is positive definite if and only if $\sigma_i > 0$ for each i.

Another important fact that we shall exploit is the Sylvester criterion for positive definite matrices. Given an $n \times n$ matrix $A = (a_{ij})$ and $k \in \{1, \dots, n\}$, let us write A_k for, the $k \times k$ matrix obtained from A by keeping the first k rows and columns and deleting the rest. For example,

$$A_1 = (a_{11}), \qquad A_2 = \begin{pmatrix} a_{11} & a_{12} \\ a_{21} & a_{22} \end{pmatrix}, \qquad A_n = A.$$

Theorem C.1.2 (Sylvester's criterion) *Let A be a hermitian $n \times n$ matrix. Then A is positive definite if and only if* $\det(A_k) > 0$ *for all $k \in \{1, \dots, n\}$.*

Proof Suppose that A is positive definite. Then all its eigenvalues are strictly positive, so $\det(A) > 0$. Also, each A_k is positive definite, and so $\det(A_k) > 0$ for all k.

Conversely, suppose that $\det(A_k) > 0$ for $1 \leq k \leq n$. By induction, the matrix A_{n-1} is positive definite. Let M be the sum of the eigenspaces corresponding to all the negative eigenvalues of A. As A_{n-1} is positive definite, we must have $M \cap (\mathbb{C}^{n-1} \times \{0\}) = \{0\}$, and consequently $\dim M \leq 1$. Therefore A has at most one negative eigenvalue. If it has exactly one negative eigenvalue, then $\det(A) < 0$, contrary to hypothesis. Therefore all the eigenvalues are positive, and A is positive definite. □

Note that having $\det(A_k) \geq 0$ for all $k \in \{1, \ldots, n\}$ does *not* imply that A is positive semi-definite. Consider, for example, the matrix

$$A := \begin{pmatrix} 0 & 0 \\ 0 & -1 \end{pmatrix}.$$

C.2 Hadamard products

Definition C.2.1 Let $A = (a_{jk})$ and $B = (b_{jk})$ be complex $n \times n$ matrices. The *Hadamard product* of A and B is the $n \times n$ matrix $A \circ B := (c_{jk})$, where $c_{jk} = a_{jk} b_{jk}$ $(j, k \in \{1, \ldots, n\})$.

Thus the Hadamard product is obtained simply by coordinatewise multiplication.

Theorem C.2.2 (Schur product theorem) *If A and B are positive semi-definite matrices, then $A \circ B$ is also positive semi-definite. If both A and B are positive definite, then so is $A \circ B$.*

Proof Consider the representation (C.2) and write $U = (u_{ij})$. Then, for each i and j,

$$a_{ij} = \sum_{k=1}^{n} \sigma_k u_{ik} \overline{u}_{jk}.$$

Therefore, for any choice of $\lambda_1, \ldots, \lambda_n \in \mathbb{C}$, we have

$$\begin{aligned} \sum_{i=1}^{n} \sum_{j=1}^{n} a_{ij} b_{ij} \lambda_i \overline{\lambda}_j &= \sum_{i=1}^{n} \sum_{j=1}^{n} \sum_{k=1}^{n} \sigma_k u_{ik} \overline{u}_{jk} b_{ij} \lambda_i \overline{\lambda}_j \\ &= \sum_{k=1}^{n} \sigma_k \Big(\sum_{i=1}^{n} \sum_{j=1}^{n} b_{ij} (\lambda_i u_{ik}) \overline{(\lambda_j u_{jk})} \Big). \end{aligned}$$

Since B is positive semi-definite, the double sum in brackets is non-negative. Also, since A is positive semi-definite, we have $\sigma_k \geq 0$ for each k. Hence

$$\sum_{i,j=1}^{n} a_{ij} b_{ij} \lambda_i \overline{\lambda}_j \geq 0.$$

Thus $A \circ B$ is positive semi-definite.

If both A and B are positive definite, then there exists $t > 0$ such that $A - tI$ and $B - tI$ are both positive semi-definite. Writing

$$A \circ B = \Big((A - tI) \circ B\Big) + \Big(tI \circ (B - tI)\Big) + t^2 I,$$

exhibits $A \circ B$ as a sum of three positive semi-definite matrices, the last of which is positive definite. It follows that $A \circ B$ is itself positive definite. □

Theorem C.2.3 (Oppenheim's inequality) *Let A and B be positive semi-definite $n \times n$ matrices. Then*

$$\det(A \circ B) \geq \det(A) \det(I \circ B).$$

Proof By Theorem C.2.2 the matrix $A \circ B$ is positive semi-definite and so $\det(A \circ B) \geq 0$. This proves the result in the case when $\det(A) = 0$, so we may as well assume that $\det(A) > 0$. This means that A is positive definite. Let J be the $n \times n$ matrix all of whose entries are zero except for the very last one, which equals 1. Then, for each $t \in \mathbb{R}$, we have

$$(A - tJ)_k = A_k \qquad (1 \leq k \leq n - 1),$$

and expansion with respect to the last row shows that

$$\det(A - tJ) = \det(A) - t \det(A_{n-1}) \qquad (t \in \mathbb{R}).$$

Thus, by Sylvester's criterion Theorem C.1.2, the matrix $A - tJ$ is positive definite provided that $t < \det(A)/\det(A_{n-1})$. For all such t, Theorem C.2.2 tells us that $(A - tJ) \circ B$ is positive semi-definite, and hence $\det((A - tJ) \circ B) \geq 0$. After expansion with respect to the bottom row, this last inequality is equivalent to

$$\det(A \circ B) \geq \det(A_{n-1} \circ B_{n-1}) t b_{nn}.$$

As this holds for $t < \det(A)/\det(A_{n-1})$, it also holds for $t = \det(A)/\det(A_{n-1})$, whence

$$\frac{\det(A \circ B)}{\det(A)} \geq \frac{\det(A_{n-1} \circ B_{n-1})}{\det(A_{n-1})} b_{nn}.$$

Iterating this gives

$$\frac{\det(A \circ B)}{\det(A)} \geq \prod_{j=1}^{n} b_{jj},$$

which is the desired inequality. □

Taking $B = I$ in this theorem, we immediately obtain the following corollary.

Corollary C.2.4 (Hadamard's inequality) *If A is an $n \times n$ positive semi-definite matrix, then*

$$\det(A) \leq \prod_{j=1}^{n} a_{jj}.$$

Remark Oppenheim's inequality can be strengthened as follows. If A and B are positive semi-definite $n \times n$ matrices, then

$$\det(A \circ B) + \det(A)\det(B) \geq \det(A)\det(I \circ B) + \det(A \circ I)\det(B).$$

This inequality is known as Schur's inequality. We do not stop to give a proof since the result is not needed in the book.

Appendix D

Regularization and the rising-sun lemma

In this appendix we prove a regularization lemma, Lemma 9.4.5, used in establishing Theorem 9.4.1. The principal tool is the notion of the increasing regularization of a function, and we begin by examining this separately.

D.1 Increasing regularization

Definition D.1.1 Given a function $u : [0, \infty) \to [0, \infty)$, we define its *increasing regularization* $\widetilde{u} : [0, \infty) \to [0, \infty)$ by

$$\widetilde{u}(x) := \inf\{u(y) : y \geq x\} \qquad (x \geq 0).$$

Clearly $\widetilde{u}$ is increasing and $\widetilde{u} \leq u$. Also, $\widetilde{u}$ is maximal with these two properties, in the sense that, if v is any increasing function with $v \leq u$, then also $v \leq \widetilde{u}$.

We shall need two results about increasing regularizations. The first is a version of the so-called rising-sun lemma of F. Riesz.

Lemma D.1.2 *Let $u : [0, \infty) \to [0, \infty)$ be a function that is lower semicontinuous and right-continuous. Let $\widetilde{u}$ be the increasing regularization of u and define $U := \{x \in [0, \infty) : \widetilde{u}(x) < u(x)\}$. Then U is open in $[0, \infty)$. Further, if a, b are the endpoints of any component of U, then $u(a) \geq u(b)$.*

Proof Let $x \in U$. Then there exists $y > x$ such that $u(y) < u(x)$. By lower semicontinuity $u(y) < u(x')$ for all x' in a neighborhood of x. All such x' also belong to U. Thus U is open in $[0, \infty)$.

Now let a, b be the endpoints of a component of U. Since U is open in $[0, \infty)$, we have $b \notin U$, and hence $u(y) \geq u(b)$ for all $y \geq b$. Let $x \in (a, b)$. As u is lower semicontinuous on the compact set $[x, b]$, its minimum on this set is attained, at x_0 say. We then have $u(y) \geq u(x_0)$ for all $y \geq x_0$, which implies that

$\widetilde{u}(x_0) = u(x_0)$ and so $x_0 \notin U$. The only possibility is that $x_0 = b$. Thus $u \geq u(b)$ on $[x, b]$, and in particular $u(x) \geq u(b)$. Finally, letting $x \to a$ and using the right-continuity of u, we obtain $u(a) \geq u(b)$. □

In the rising-sun terminology, the set U corresponds to the shade. The second result that we need, taken from [44], is an estimate for the proportion of $[0, \infty)$ that stays in the sun. Recall that the *lower density* of a Borel set $B \subset [0, \infty)$ is defined by

$$\rho_-(B) := \liminf_{x \to \infty} \frac{|B \cap [0, x]|}{x}.$$

Lemma D.1.3 *Let $u : [0, \infty) \to [0, \infty)$ be a positive function such that $u(x) - x$ is decreasing. Set $S := \{x \in [0, \infty) : \widetilde{u}(x) = u(x)\}$. Then S is a Borel set and*

$$\rho_-(S) \geq \liminf_{x \to \infty} \frac{u(x)}{x}.$$

Proof As $u(x) - x$ is decreasing, it follows that $u_1(x) := \lim_{y \downarrow x} u(y)$ exists for all x. The function u_1 is both lower semicontinuous and right-continuous, and $u_1(x) - x$ is decreasing. Further, we have both $u_1 = u$ and $\widetilde{u}_1 = \widetilde{u}$ except on countable sets. Thus, we may as well suppose from the outset that u is lower semicontinuous and right-continuous, so that Lemma D.1.2 applies.

We may also suppose that $u(x) \to \infty$ as $x \to \infty$ for, if not, then necessarily $\liminf_{x \to \infty} u(x)/x = 0$, and there is nothing to prove. As a consequence of this supposition, S is necessarily unbounded.

Let $y \in S$. Let $I_1, \dots, I_n$ be a finite set of components of $U := [0, \infty) \setminus S$ lying in $[0, y]$. Let a_j, b_j be the endpoints of I_j. Renumbering, if necessary, we may suppose that $0 \leq a_1 < b_1 < \cdots < a_n < b_n \leq y$. Then

$$\begin{aligned}|I_1 \cup \cdots \cup I_n| &= \sum_{j=1}^{n} (b_j - a_j) \\ &\leq \sum_{j=1}^{n} (b_j - u(b_j) - a_j + u(a_j)) \\ &\leq y - u(y) + u(0),\end{aligned}$$

where, for the first inequality we used Lemma D.1.2, and for the second the fact that $u(x) - x$ is decreasing. As this holds for any such set of components, it follows that $|U \cap [0, y]| \leq y - u(y) + u(0)$. Recalling that U is the complement of S, we deduce that

$$|S \cap [0, y]| \geq u(y) - u(0) \quad (y \in S).$$

Now, given $x \in [0, \infty)$, let y be the smallest element of S such that $y \geq x$.

Then

$$\frac{|S \cap [0,x]|}{x} = \frac{|S \cap [0,y]|}{x} \geq \frac{|S \cap [0,y]|}{y} \geq \frac{u(y) - u(0)}{y}.$$

It follows that $\liminf_{x\to\infty} |S \cap [0,x]|/x \geq \liminf_{y\to\infty} u(y)/y$, which is what we had to prove. □

D.2 Proof of the regularization lemma

For convenience, we re-state the regularization lemma, Lemma 9.4.5.

Lemma D.2.1 *Let $a > 0$, let $\beta \in (0,1]$ and let $\phi : (0,a] \to (0,\infty)$ be a function such that*

- *$\phi(t)/t$ is decreasing,*
- *$0 < \phi(t) \leq t^\beta$ for all $t \in (0,a]$,*
- *$\int_0^a dt/\phi(t) = \infty$.*

Then, given $\alpha \in (0,\beta)$, there exists a function $\psi : (0,a] \to (0,\infty)$ such that

- *$\psi(t)/t^\alpha$ is increasing,*
- *$\phi(t) \leq \psi(t) \leq t^\beta$ for all $t \in (0,a]$,*
- *$\int_0^a dt/\psi(t) = \infty$.*

The proof is based on increasing regularization, as developed in the previous section, and on the following simple result about sets of positive lower density.

Lemma D.2.2 *Let $v : [0,\infty) \to [0,\infty)$ be a positive decreasing function such that $\int_0^\infty v(x)\,dx = \infty$. Then $\int_B v(x)\,dx = \infty$ for every Borel set $B \subset [0,\infty)$ with $\rho_-(B) > 0$.*

Proof Suppose that $\rho_-(B) > 0$. Then there exists $\lambda > 0$ such that, for all sufficiently large x,

$$|B \cap [0,x]| \geq \lambda x.$$

Fix $a > 1/\lambda$. Then, for all sufficiently large k,

$$\begin{aligned}\int_{B\cap[a^{k-1},a^k]} v(x)\,dx &\geq v(a^k)|B \cap [a^{k-1},a^k]| \\ &\geq v(a^k)(|B \cap [0,a^k]| - a^{k-1}) \\ &\geq v(a^k)(\lambda a^k - a^{k-1}).\end{aligned}$$

Also, for all k,

$$\int_{[a^k,a^{k+1}]} v(x)\,dx \le v(a^k)(a^{k+1} - a^k).$$

Hence, for all sufficiently large k,

$$\int_{B\cap[a^{k-1},a^k]} v(x)\,dx \ge \frac{\lambda - 1/a}{a-1} \int_{[a^k,a^{k+1}]} v(x)\,dx.$$

Summing over these k, we deduce that $\int_B v(x)\,dx = \infty$. □

Proof of Lemma D.2.1 By a simple change of scale, we can reduce to the case where $a = 1$. This will simplify the notation in what follows.

Define $u : [0, \infty) \to [0, \infty)$ by the formula

$$u(x) := -\frac{1}{1-\alpha} \log \phi(e^{-x}) - \frac{\alpha}{1-\alpha} x \qquad (x \ge 0).$$

The properties of ϕ are reflected in u as follows:

$$\phi(t)/t \text{ decreasing} \iff u(x) - x \text{ decreasing},$$

$$\phi(t) \le t^\beta \iff u(x) \ge \frac{\beta - \alpha}{1-\alpha} x,$$

$$\int_0^1 \frac{dt}{\phi(t)} = \infty \iff \int_0^\infty e^{(1-\alpha)(u(x)-x)}\,dx = \infty.$$

Let $\widetilde{u} : [0, \infty) \to [0, \infty)$ be the increasing regularization of u, and let us define $\psi : (0, 1] \to [0, \infty)$ to be the function satisfying the formula

$$\widetilde{u}(x) = -\frac{1}{1-\alpha} \log \psi(e^{-x}) - \frac{\alpha}{1-\alpha} x \qquad (x \ge 0).$$

The desired properties of ψ correspond to properties of $\widetilde{u}$ as follows:

$$\psi(t)/t^\alpha \text{ increasing} \iff \widetilde{u}(x) \text{ increasing},$$

$$\phi(t) \le \psi(t) \le t^\beta \iff u(x) \ge \widetilde{u}(x) \ge \frac{\beta-\alpha}{1-\alpha} x,$$

$$\int_0^1 \frac{dt}{\psi(t)} = \infty \iff \int_0^\infty e^{(1-\alpha)(\widetilde{u}(x)-x)}\,dx = \infty.$$

It thus suffices to prove these three properties of $\widetilde{u}$. The first two are obvious. For the third, we remark that, writing $S := \{x \in [0, \infty) : \widetilde{u}(x) = u(x)\}$, we have

$$\int_0^\infty e^{(1-\alpha)(\widetilde{u}(x)-x)}\,dx \ge \int_S e^{(1-\alpha)(\widetilde{u}(x)-x)}\,dx = \int_S e^{(1-\alpha)(u(x)-x)}\,dx.$$

Also $e^{(1-\alpha)(\widetilde{u}(x)-x)}$ is a decreasing function and, by Lemma D.1.3, we have $\rho_-(S) \ge (\beta - \alpha)/(\alpha - 1) > 0$. Therefore, by Lemma D.2.2, the last integral diverges, and the proof is complete. □

References

[1] Adams, D. R., and Hedberg, L. I. 1996. *Function Spaces and Potential Theory*. Grundlehren der Mathematischen Wissenschaften, vol. 314. Berlin: Springer-Verlag.

[2] Agler, J. 1988. *Some interpolation theorems of Nevanlinna–Pick type*. Unpublished manuscript.

[3] Agler, J. 1990. A disconjugacy theorem for Toeplitz operators. *Amer. J. Math.*, **112**(1), 1–14.

[4] Agler, J., and McCarthy, J. E. 2002. *Pick Interpolation and Hilbert Function Spaces*. Graduate Studies in Mathematics, vol. 44. Providence, RI: American Mathematical Society.

[5] Ahlfors, L. V. 1973. *Conformal Invariants: Topics in Geometric Function Theory*. McGraw-Hill Series in Higher Mathematics. New York: McGraw-Hill.

[6] Aikawa, H., and Essén, M. 1996. *Potential Theory—Selected Topics*. Lecture Notes in Mathematics, vol. 1633. Berlin: Springer-Verlag.

[7] Aleman, A. 1993. *The multiplication operator on Hilbert spaces of analytic functions*. Habilitationsschrift, Fern Universität, Hagen.

[8] Arazy, J., and Fisher, S. D. 1985. The uniqueness of the Dirichlet space among Möbius-invariant Hilbert spaces. *Illinois J. Math.*, **29**(3), 449–462.

[9] Arcozzi, N., Rochberg, R., and Sawyer, E. T. 2002. Carleson measures for analytic Besov spaces. *Rev. Mat. Iberoamericana*, **18**(2), 443–510.

[10] Arcozzi, N., Rochberg, R., and Sawyer, E. T. 2008. Carleson measures for the Drury-Arveson Hardy space and other Besov-Sobolev spaces on complex balls. *Adv. Math.*, **218**(4), 1107–1180.

[11] Arcozzi, N., Rochberg, R., Sawyer, E. T., and Wick, B. D. 2011. The Dirichlet space: a survey. *New York J. Math.*, **17A**, 45–86.

[12] Armitage, D. H., and Gardiner, S. J. 2001. *Classical Potential Theory*. Springer Monographs in Mathematics. London: Springer-Verlag.

[13] Bénéteau, C., Condori, A., Liaw, C., Seco, D., and Sola, A. Cyclicity in Dirichlet-type spaces and extremal polynomials. *J. Anal. Math.* To appear.

[14] Berenstein, C. A., and Gay, R. 1991. *Complex Variables*. Graduate Texts in Mathematics, vol. 125. New York: Springer-Verlag.

[15] Beurling, A. 1933. *Études sur un problème de majoration*. Doctoral thesis, Uppsala University.

[16] Beurling, A. 1940. Ensembles exceptionnels. *Acta Math.*, **72**, 1–13.

[17] Bishop, C. J. 1994. *Interpolating sequences for the Dirichlet space and its multipliers*. Unpublished manuscript.

[18] Bøe, B. 2005. An interpolation theorem for Hilbert spaces with Nevanlinna-Pick kernel. *Proc. Amer. Math. Soc.*, **133**(7), 2077–2081.

[19] Bogdan, K. 1996. On the zeros of functions with finite Dirichlet integral. *Kodai Math. J.*, **19**(1), 7–16.

[20] Borichev, A. 1993. *A note on Dirichlet-type spaces*. Uppsala University Department of Mathematics Report 11.

[21] Borichev, A. 1994. *Boundary behavior in Dirichlet-type spaces*. Uppsala University Department of Mathematics Report 3.

[22] Bourdon, P. S. 1986. Cellular-indecomposable operators and Beurling's theorem. *Michigan Math. J.*, **33**(2), 187–193.

[23] Brown, L., and Cohn, W. 1985. Some examples of cyclic vectors in the Dirichlet space. *Proc. Amer. Math. Soc.*, **95**(1), 42–46.

[24] Brown, L., and Shields, A. L. 1984. Cyclic vectors in the Dirichlet space. *Trans. Amer. Math. Soc.*, **285**(1), 269–303.

[25] Carleson, L. 1952a. On the zeros of functions with bounded Dirichlet integrals. *Math. Z.*, **56**, 289–295.

[26] Carleson, L. 1952b. Sets of uniqueness for functions regular in the unit circle. *Acta Math.*, **87**, 325–345.

[27] Carleson, L. 1960. A representation formula for the Dirichlet integral. *Math. Z.*, **73**, 190–196.

[28] Carleson, L. 1962. Interpolations by bounded analytic functions and the corona problem. *Ann. Math. (2)*, **76**, 547–559.

[29] Carleson, L. 1967. *Selected Problems on Exceptional Sets*. Van Nostrand Mathematical Studies, No. 13. Princeton, NJ: Van Nostrand.

[30] Carlsson, M. 2008. On the Cowen-Douglas class for Banach space operators. *Integral Equations Operator Theory*, **61**(4), 593–598.

[31] Caughran, J. G. 1969. Two results concerning the zeros of functions with finite Dirichlet integral. *Canad. J. Math.*, **21**, 312–316.

[32] Caughran, J. G. 1970. Zeros of analytic functions with infinitely differentiable boundary values. *Proc. Amer. Math. Soc.*, **24**, 700–704.

[33] Chacón, G. R. 2011. Carleson measures on Dirichlet-type spaces. *Proc. Amer. Math. Soc.*, **139**(5), 1605–1615.

[34] Chang, S.-Y. A., and Marshall, D. E. 1985. On a sharp inequality concerning the Dirichlet integral. *Amer. J. Math.*, **107**(5), 1015–1033.

[35] Chartrand, R. 2002. Toeplitz operators on Dirichlet-type spaces. *J. Operator Theory*, **48**(1), 3–13.

[36] Cowen, C. C., and MacCluer, B. D. 1995. *Composition Operators on Spaces of Analytic Functions*. Studies in Advanced Mathematics. Boca Raton, FL: CRC Press.

[37] Cowen, M. J., and Douglas, R. G. 1978. Complex geometry and operator theory. *Acta Math.*, **141**(3–4), 187–261.

[38] Doob, J. L. 1984. *Classical Potential Theory and its Probabilistic Counterpart*. Grundlehren der Mathematischen Wissenschaften, vol. 262. New York: Springer-Verlag.

[39] Douglas, J. 1931. Solution of the problem of Plateau. *Trans. Amer. Math. Soc.*, **33**(1), 263–321.

[40] Duren, P. L. 1970. *Theory of H^p Spaces*. Pure and Applied Mathematics, Vol. 38. New York: Academic Press.

[41] Dyn′kin, E. M. 1972. *Extensions and integral representations of smooth functions of one complex variable*. Dissertation, Leningrad.

[42] El-Fallah, O., Kellay, K., Mashreghi, J., and Ransford, T. One-box conditions for Carleson measures for the Dirichlet space. *Proc. Amer. Math. Soc.* To appear.

[43] El-Fallah, O., Kellay, K., and Ransford, T. 2006. Cyclicity in the Dirichlet space. *Ark. Mat.*, **44**(1), 61–86.

[44] El-Fallah, O., Kellay, K., and Ransford, T. 2009. On the Brown-Shields conjecture for cyclicity in the Dirichlet space. *Adv. Math.*, **222**(6), 2196–2214.

[45] El-Fallah, O., Kellay, K., Shabankhah, M., and Youssfi, H. 2011. Level sets and composition operators on the Dirichlet space. *J. Funct. Anal.*, **260**(6), 1721–1733.

[46] El-Fallah, O., Kellay, K., Mashreghi, J., and Ransford, T. 2012. A self-contained proof of the strong-type capacitary inequality for the Dirichlet space. Pages 1–20 of: *Complex Analysis and Potential Theory*. CRM Proc. Lecture Notes, vol. 55. Providence, RI: American Mathematical Society.

[47] Essén, M. 1987. Sharp estimates of uniform harmonic majorants in the plane. *Ark. Mat.*, **25**(1), 15–28.

[48] Frostman, O. 1935. Potentiel d'équilibre et capacité des ensembles avec quelques applications à la théorie des fonctions. Thesis. *Meddel. Lunds Univ. Mat. Sem.*, **3**, 1–118.

[49] Gallardo-Gutiérrez, E. A., and González, M. J. 2003. Exceptional sets and Hilbert-Schmidt composition operators. *J. Funct. Anal.*, **199**(2), 287–300.

[50] Garnett, J. B. 2007. *Bounded Analytic Functions*. revised first edn. Graduate Texts in Mathematics, vol. 236. New York: Springer.

[51] Garnett, J. B., and Marshall, D. E. 2005. *Harmonic Measure*. New Mathematical Monographs, vol. 2. Cambridge: Cambridge University Press.

[52] Guillot, D. 2012a. Blaschke condition and zero sets in weighted Dirichlet spaces. *Ark. Mat.*, **50**(2), 269–278.

[53] Guillot, D. 2012b. Fine boundary behavior and invariant subspaces of harmonically weighted Dirichlet spaces. *Complex Anal. Operator Theory*, **6**(6), 1211–1230.

[54] Hansson, K. 1979. Imbedding theorems of Sobolev type in potential theory. *Math. Scand.*, **45**(1), 77–102.

[55] Hardy, G. H. 1949. *Divergent Series*. Oxford: Clarendon Press.

[56] Hastings, W. W. 1975. A Carleson measure theorem for Bergman spaces. *Proc. Amer. Math. Soc.*, **52**, 237–241.

[57] Hayman, W. K., and Kennedy, P. B. 1976. *Subharmonic Functions,* vol. 1. London Mathematical Society Monographs, No. 9. London: Academic Press [Harcourt Brace Jovanovich Publishers].

[58] Hedenmalm, H., and Shields, A. L. 1990. Invariant subspaces in Banach spaces of analytic functions. *Michigan Math. J.*, **37**(1), 91–104.

[59] Helms, L. L. 1975. *Introduction to Potential Theory*. Pure and Applied Mathematics, vol. 22. Huntington, NY: Robert E. Krieger.

[60] Hille, E. 1962. *Analytic Function Theory,* vol. II. Introductions to Higher Mathematics. Boston, MA: Ginn and Co.

[61] Kahane, J.-P., and Salem, R. 1994. *Ensembles parfaits et séries trigonométriques.* Second edn. Paris: Hermann.

[62] Kellay, K. 2011. Poincaré type inequality for Dirichlet spaces and application to the uniqueness set. *Math. Scand.*, **108**(1), 103–114.

[63] Kellay, K., and Mashreghi, J. 2012. On zero sets in the Dirichlet space. *J. Geom. Anal.*, **22**(4), 1055–1070.

[64] Kerman, R., and Sawyer, E. T. 1988. Carleson measures and multipliers of Dirichlet-type spaces. *Trans. Amer. Math. Soc.*, **309**(1), 87–98.

[65] Koosis, P. 1992. *The Logarithmic Integral II.* Cambridge Studies in Advanced Mathematics, vol. 21. Cambridge: Cambridge University Press.

[66] Koosis, P. 1998. *Introduction to H_p Spaces.* Second edn. Cambridge Tracts in Mathematics, vol. 115. Cambridge: Cambridge University Press.

[67] Korenblum, B. I. 1972. Invariant subspaces of the shift operator in a weighted Hilbert space. *Math. USSR-Sb.*, **18**, 111–138.

[68] Korenblum, B. I. 2006. Blaschke sets for Bergman spaces. Pages 145–152 of: *Bergman Spaces and Related Topics in Complex Analysis.* Contemp. Math., vol. 404. Providence, RI: American Mathematical Society.

[69] Korolevič, V. S. 1970. A certain theorem of Beurling and Carleson. *Ukrainian Math. J.*, **22**(6), 710–714.

[70] Landkof, N. S. 1972. *Foundations of Modern Potential Theory.* Grundlehren der mathematischen Wissenschaften, vol. 180. New York: Springer-Verlag.

[71] Lefèvre, P., Li, D., Queffélec, H., and Rodríguez-Piazza, L. 2013. Compact composition operators on the Dirichlet space and capacity of sets of contact points. *J. Funct. Anal.*, **264**(4), 895–919.

[72] Luecking, D. H. 1987. Trace ideal criteria for Toeplitz operators. *J. Funct. Anal.*, **73**(2), 345–368.

[73] MacCluer, B. D., and Shapiro, J. H. 1986. Angular derivatives and compact composition operators on the Hardy and Bergman spaces. *Canad. J. Math.*, **38**(4), 878–906.

[74] Malliavin, P. 1977. Sur l'analyse harmonique de certaines classes de séries de Taylor. Pages 71–91 of: *Symposia Mathematica, Vol. XXII (Convegno sull'Analisi Armonica e Spazi di Funzioni su Gruppi Localmente Compatti, INDAM, Rome, 1976).* London: Academic Press.

[75] Marshall, D. E. 1989. A new proof of a sharp inequality concerning the Dirichlet integral. *Ark. Mat.*, **27**(1), 131–137.

[76] Marshall, D. E., and Sundberg, C. 1989. *Interpolating sequences for the multipliers of the Dirichlet space.* Unpublished manuscript.

[77] Mashreghi, J. 2009. *Representation Theorems in Hardy Spaces.* London Mathematical Society Student Texts, vol. 74. Cambridge: Cambridge University Press.

[78] Mashreghi, J., Ransford, T., and Shabankhah, M. 2010. Arguments of zero sets in the Dirichlet space. Pages 143–148 of: *Hilbert Spaces of Analytic Functions.* CRM Proc. Lecture Notes, vol. 51. Providence, RI: American Mathematical Society.

[79] Maz'ya, V. G., and Havin, V. P. 1973. Application of the (p, l)-capacity to certain problems of the theory of exceptional sets. *Mat. Sb. (N.S.)*, **90(132)**, 558–591, 640.

[80] Meyers, N. G. 1970. A theory of capacities for potentials of functions in Lebesgue classes. *Math. Scand.*, **26**, 255–292 (1971).

[81] Monterie, M. A. 1997. Capacities of certain Cantor sets. *Indag. Math. (N.S.)*, **8**(2), 247–266.

[82] Nagel, A., and Stein, E. M. 1984. On certain maximal functions and approach regions. *Adv. Math.*, **54**(1), 83–106.

[83] Nagel, A., Rudin, W., and Shapiro, J. H. 1982. Tangential boundary behavior of functions in Dirichlet-type spaces. *Ann. Math. (2)*, **116**(2), 331–360.

[84] Nagel, A., Rudin, W., and Shapiro, J. H. 1983. Tangential boundary behavior of harmonic extensions of L^p potentials. Pages 533–548 of: *Conference on Harmonic Analysis in Honor of Antoni Zygmund, Vol. I, II (Chicago, IL, 1981)*. Wadsworth Math. Ser. Belmont, CA: Wadsworth.

[85] Nehari, Z. 1975. *Conformal Mapping*. New York: Dover Publications.

[86] Novinger, W. P. 1971. Holomorphic functions with infinitely differentiable boundary values. *Illinois J. Math.*, **15**, 80–90.

[87] Ohtsuka, M. 1957. Capacité d'ensembles de Cantor généralisés. *Nagoya Math. J.*, **11**, 151–160.

[88] Olin, R. F., and Thomson, J. E. 1984. Cellular-indecomposable subnormal operators. *Integral Equations Operator Theory*, **7**(3), 392–430.

[89] Parrott, S. 1978. On a quotient norm and the Sz.-Nagy–Foiaş lifting theorem. *J. Funct. Anal.*, **30**(3), 311–328.

[90] Peller, V. V., and Hruščev, S. V. 1982. Hankel operators, best approximations and stationary Gaussian processes. *Russian Math. Surveys*, **37**(1), 61–144.

[91] Pommerenke, Ch. 1975. *Univalent Functions*. Göttingen: Vandenhoeck & Ruprecht.

[92] Pommerenke, Ch. 1992. *Boundary Behaviour of Conformal Maps*. Grundlehren der Mathematischen Wissenschaften, vol. 299. Berlin: Springer-Verlag.

[93] Ransford, T. 1995. *Potential Theory in the Complex Plane*. London Mathematical Society Student Texts, vol. 28. Cambridge: Cambridge University Press.

[94] Ransford, T., and Selezneff, A. 2012. Capacity and covering numbers. *Potential Anal.*, **36**, 223–233.

[95] Richter, S. 1988. Invariant subspaces of the Dirichlet shift. *J. Reine Angew. Math.*, **386**, 205–220.

[96] Richter, S. 1991. A representation theorem for cyclic analytic two-isometries. *Trans. Amer. Math. Soc.*, **328**(1), 325–349.

[97] Richter, S., and Shields, A. L. 1988. Bounded analytic functions in the Dirichlet space. *Math. Z.*, **198**(2), 151–159.

[98] Richter, S., and Sundberg, C. 1991. A formula for the local Dirichlet integral. *Michigan Math. J.*, **38**(3), 355–379.

[99] Richter, S., and Sundberg, C. 1992. Multipliers and invariant subspaces in the Dirichlet space. *J. Operator Theory*, **28**(1), 167–186.

[100] Richter, S., and Sundberg, C. 1994. Invariant subspaces of the Dirichlet shift and pseudocontinuations. *Trans. Amer. Math. Soc.*, **341**(2), 863–879.

[101] Richter, S., Ross, W. T., and Sundberg, C. 2004. Zeros of functions with finite Dirichlet integral. *Proc. Amer. Math. Soc.*, **132**(8), 2361–2365.

[102] Rochberg, R., and Wu, Z. J. 1992. Toeplitz operators on Dirichlet spaces. *Integral Equations Operator Theory*, **15**(2), 325–342.

[103] Ross, W. T. 2006. The classical Dirichlet space. Pages 171–197 of: *Recent Advances in Operator-Related Function Theory*. Contemp. Math., vol. 393. Providence, RI: American Mathematical Society.

[104] Rudin, W. 1987. *Real and Complex Analysis*. Third edn. New York: McGraw-Hill.

[105] Rudin, W. 1992. Power series with zero-sum on countable sets. *Complex Variables Theory Appl.*, **18**(3–4), 283–284.

[106] Saff, E. B., and Totik, V. 1997. *Logarithmic Potentials with External Fields*. Grundlehren der Mathematischen Wissenschaften, vol. 316. Berlin: Springer-Verlag.

[107] Sarason, D. 1986. Doubly shift-invariant spaces in H^2. *J. Operator Theory*, **16**(1), 75–97.

[108] Sarason, D. 1997. Local Dirichlet spaces as de Branges-Rovnyak spaces. *Proc. Amer. Math. Soc.*, **125**(7), 2133–2139.

[109] Seip, K. 2004. *Interpolation and Sampling in Spaces of Analytic Functions*. University Lecture Series, vol. 33. Providence, RI: American Mathematical Society.

[110] Shapiro, H. S., and Shields, A. L. 1962. On the zeros of functions with finite Dirichlet integral and some related function spaces. *Math. Z.*, **80**, 217–229.

[111] Shapiro, J. H. 1980. Cauchy transforms and Beurling-Carleson-Hayman thin sets. *Michigan Math. J.*, **27**(3), 339–351.

[112] Shapiro, J. H. 1987. The essential norm of a composition operator. *Ann. Math. (2)*, **125**(2), 375–404.

[113] Shapiro, J. H. 1993. *Composition Operators and Classical Function Theory*. Universitext: Tracts in Mathematics. New York: Springer-Verlag.

[114] Shields, A. L. 1983. An analogue of the Fejér-Riesz theorem for the Dirichlet space. Pages 810–820 of: *Conference on Harmonic Analysis in Honor of Antoni Zygmund, Vol. I, II (Chicago, IL, 1981)*. Wadsworth Math. Ser. Belmont, CA: Wadsworth.

[115] Shimorin, S. M. 1998. Reproducing kernels and extremal functions in Dirichlet-type spaces. *Zap. Nauchn. Sem. S.-Peterburg. Otdel. Mat. Inst. Steklov. (POMI)*, **255** (Issled. po Linein. Oper. i Teor. Funkts. 26), 198–220, 254. Translation in *J. Math. Sci. (New York)* **107**(4) (2001), 4108–4124.

[116] Shimorin, S. M. 2002. Complete Nevanlinna-Pick property of Dirichlet-type spaces. *J. Funct. Anal.*, **191**(2), 276–296.

[117] Stegenga, D. A. 1980. Multipliers of the Dirichlet space. *Illinois J. Math.*, **24**(1), 113–139.

[118] Taylor, B. A., and Williams, D. L. 1970. Ideals in rings of analytic functions with smooth boundary values. *Canad. J. Math.*, **22**, 1266–1283.

[119] Taylor, G. D. 1966. Multipliers on D_α. *Trans. Amer. Math. Soc.*, **123**, 229–240.

[120] Tsuji, M. 1975. *Potential Theory in Modern Function Theory*. New York: Chelsea.

[121] Twomey, J. B. 1989. Tangential limits for certain classes of analytic functions. *Mathematika*, **36**(1), 39–49.

[122] Twomey, J. B. 2002. Tangential boundary behaviour of harmonic and holomorphic functions. *J. London Math. Soc. (2)*, **65**(1), 68–84.
[123] Wojtaszczyk, P. 1991. *Banach Spaces for Analysts*. Cambridge Studies in Advanced Mathematics, vol. 25. Cambridge: Cambridge University Press.
[124] Wynn, A. 2011. Sufficient conditions for weighted admissibility of operators with applications to Carleson measures and multipliers. *Q. J. Math.*, **62**(3), 747–770.

Index of notation

Index

Printed in the United States
By Bookmasters